1910

Catalog of Staking Tools

MANUFACTURED BY

KENDRICK & DAVIS CO.

F. B. KENDRICK
President

R. R. WOOD
Treasurer

W. H. CROZIER
Superintendent

TRADE MARK · REG. U.S. PAT. OFFICE

Lebanon, New Hampshire

This edition, published in 2002 by Arlington Book Company Inc., is an unabridge republication of the work originally published in 1910 by Kendrick and Davis Company, Lebanon, New Hampshire, reprinted in April 1984 by K & D Manufacturing Corp., Lebanon, New Hampshire, U.S.A. under the title "Staking Tools and How to Use Them".

Printed in the United States of America by

Arlington Book Company, Inc.
2706 Elsmore Street
Fairfax, Virginia 22031-1409 USA
Email: **tranduyly@aol.com**

www.arlingtonbooks.com

ISBN 0-930163-16-8

Looking Backward

It would be interesting, did space permit, to trace carefully the development of the staking tool. We must, however, content ourselves with a very brief consideration of it, from an historical standpoint—the *present* is what most concerns us.

The earliest stakes of which we have certain knowledge were adapted to be held in the bench vise; the punches being held by hand. Such a stake is shown in Fig. 1. The punches for use with such stakes were usually put up in wooden boxes, and were commonly nothing but iron, thinly case-hardened.

Fig. 1

These were the commercial articles; makers of watches, and the expert workmen, undoubtedly had tools of their own make, of better quality.

When we examine some of the old watches, and note the workmanlike riveting and exquisite finish, we pay generous tribute of admiration to the skill of him who could, with such primitive means, produce such beautiful results; and comparing our facilities with *his*, we should resolve to produce only high class work.

The *Modern* staking tool, with revolving die, was invented by Jonas G. Hall, early in the '70's.

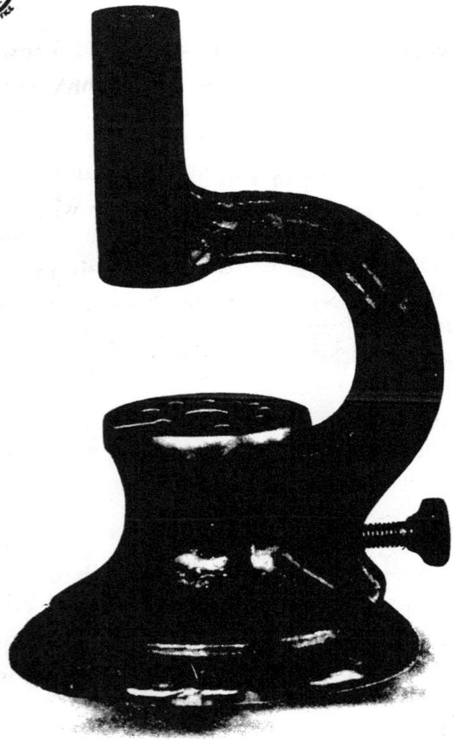

Fig 2

Fig. 2 shows his first commercial product; this tool is stamped No. 1. Tools of substantially this style were manufactured by him, at Roxbury, Vt., for many years; the only changes made being some refinements in the matter of shape, a cam die binder, operated by lever, and knurling the edge of the die. Fig. 3 shows one of the latest tools made by Mr. Hall; in point of workmanship they were of high quality.

Fig 3

On the death of Mr. Hall, the business passed into other hands, and was, within a comparatively short time, discontinued.

Meantime Mr. W. H. Crozier, who had been for many years with Mr. Hall as Superintendent, entered the employ of Kendrick & Davis, who soon began the manufacture of staking tools, under Mr. Crozier's supervision. It will be seen that he brought all the knowledge and experience, acquired in his years of service with Mr. Hall, to bear on the production of the Kendrick & Davis staking tool; this, with the superior manufacturing facilities afforded, made the production of high quality tools a certainty.

Mr. Kendrick, himself a practical watchmaker, who keeps in touch with leading men of the craft, has been quick to realize the need of improvements, and to recognize them when made. The result has been: continuous improvement in staking tools, and machinery for their production.

Many improvements were made in the original Hall tool; the weight was greatly increased, making it more efficient in such work as closing holes, etc., as well as much more convenient for all work. The number of holes in the die was increased, and many punches added to the assortment. All this, while real improvement, was not in the nature of invention.

The first real invention in Staking Tools, after Hall's original conception, was the K. & D. friction punch sustaining sleeve; this added greatly to convenience and rapidity in using the Staking Tool. When this was adopted, the shape of the frame was changed, and improvements made in the die binding mechanism; the chief feature being an adjustable step in the die bolt, to engage the binding cam.

The line of improvement culminates in INVERTO, the most important invention yet made in Staking Tools—fully described in following pages; also the new On or Off punch sustaining friction device, Micrometer Stop, and other up-to-date products.

During the closing months of 1909, Kendrick & Davis became an incorporated company; the corporate name being Kendrick & Davis Co. This change gives increased manufacturing facilities, made necessary by the rapidly increasing demand for K. & D. products.

We are including in this pamphlet the first pages of the 1910 catalog of staking tools. This will give you a brief history of the development of the staking tool set. Pages relating to discontinued tools have been eliminated and therefore page numbers have no meaning.

Roger S. Wood, G.M. & Treasurer
K & D Manufacturing Corp.
Lebanon, N.H. 03766

March 1984

Reproduced with permission from

K & D Manufacturing Corp.
Box 521
Lebanon, New Hampshire 03766 USA

STAKING TOOLS

AND HOW

—TO—

USE THEM

Wherefore

VOLUMES have been written on the Watchmakers' Lathe and its use, while the Staking Tool—next in importance to the lathe—has received but scant notice. There exists real need of a Hand-Book on the Staking Tool; to, in a measure, supply this need, is the chief object of this work. We trust that it may prove helpful to many.

In describing some of the uses of the Staking Tool, we have introduced certain small special tools, which greatly facilitate the work.

To do work *quickly* and *well*, is *important* to the modern watchmaker—if he would *succeed*.

F. R. C., February, 1910.

Contents

First Principles

THE Staking Tool is comparatively simple, and very simple to use; yet, there are *wrong ways* to use even simple tools. Note the cut, Fig. 1, which shows how a punch should be held *in all riveting operations*, hole-closing, and all work requiring a heavy blow.

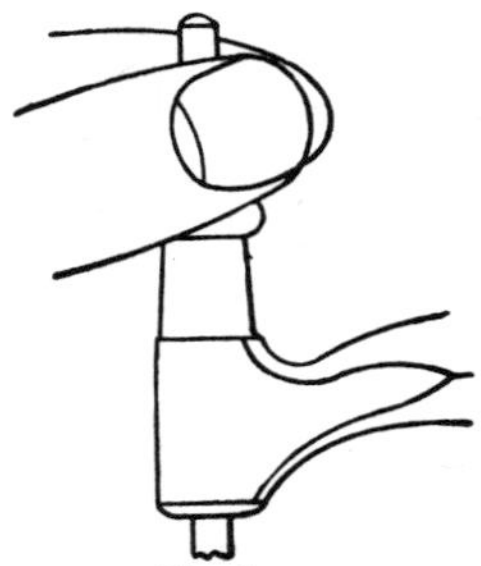

Fig. 1

As shown, the finger and thumb rest on or over the top of the staking tool, lightly grasping it, and also the punch; a slight *downward* pressure should be exerted on the punch.

By this method there is no rebound of the punch when struck; while striking a *free* punch often results in injury to the work.

As we all know, steel is a resilient material; under impact—like the blow of a hammer—a punch suddenly shortens, and as suddenly recovers; the sudden return to normal dimensions is what causes the rebound, when the punch is left free.

It is also worthy of note that the greater the inertia of the anvil—in other words, the heavier and more unyielding it is—the greater is the rebound.

As a homely illustration, throw a ball against a brick wall, and then against a light wooden one, and note the difference in the rebound.

In some exceedingly light work, such, for example, as closing a light fork, or flattening a small drill, it is not so essential to hold the punch.

As To Fitting of Dies.

Some watchmakers imagine that a staking tool die should have absolutely *no shake* on the frame, when *not bound*. To make staking tools to meet such requirements would necessitate *grinding every hole to size*, and circle, after the die was *hardened*, thus *greatly increasing the cost*, and *adding nothing* to its value.

The conditions required are: every hole should center properly with the set punch, and when the die is *bound*, the hole should *remain* accurately in position.

DON'T.

If you would have the holes in your staking tool die remain in *good condition*, sharp and smooth for fine work on watches, don't punch main springs on it. If you wish to punch main springs or do similar work, make a stump with hole of the required size; this, when dulled or damaged, can easily be repaired.

NOTE THIS.

In the following articles it has been the aim of the writer to illustrate principles, rather than to compile a lengthy catalog of operations, which are substantially alike; for example: after we have closed small holes, like train pivot holes, and large holes, like barrel holes, it is hardly necessary for us to go into details of closing screw holes, etc. The same applies to all other operations. If the workman knows how to stake in and rivet a 4th pinion, he will have no difficulty in doing the same with a third pinion.

The reader will find names and descriptions of the different punches and stumps on pages 28, 29 and 30 of the catalog section.

Punches

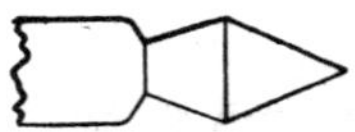

SET PUNCH. Should be used only for centering the die; any other use would soon ruin it, the point being fine and sharp. Price, 30c

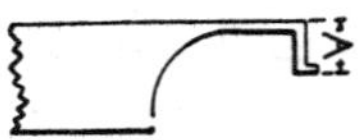

CYLINDER PUNCHES. These punches are formed by milling from the solid; they are much more rigid than the so-called forged cylinder punches, as may easily be proven by trial. Price, 30c

PUNCH FOR STRETCHING BALANCE ARMS. The balance arm should be placed on a large flat-face stump in the staking tool. Place the punch on the balance arm, close to the rim and parallel with it; a blow on the punch causes elongation of the arm, and also a slight indentation therein. Should not be used on fine balances. Proceed carefully; better practice on old balances, to train judgment as to force of blow required. Price, 25c

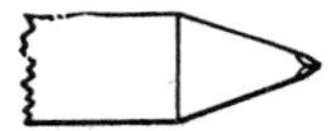

TRIANGULAR POINT. For tightening roller tables; used in the same manner as a round-faced punch. It raises three slight burrs equidistant about the hole. A round-face punch is likely to split a hard roller if one attempts to close the hole with it. Price, 25c

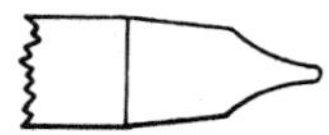

PEENING PUNCHES. Generally used in conjunction with a solid flat-faced stump, for stretching, forging, etc. Price, 25c

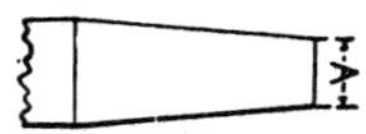

FLAT-FACE SOLID. Used for closing holes, end-shaking trains, and a great variety of work. Price, 20c

ROUND-FACE SOLID PUNCHES. For closing pivot holes, etc., generally used in conjunction with a flat-face solid stump. The high polish of our punches leaves a fine finish in the oil sinks of the holes closed. Price, 20c

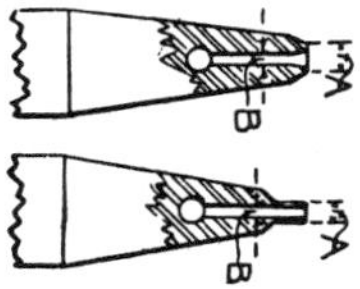

Cross Hole Punches. For driving fine staffs, arbors, pinions, etc.

Price, **30c**

Prick Punch. For fine center pricking. When it is desired to drill a hole exactly on a line, or at the intersection of lines, as, for example, where a depth is laid off by the depthing tool, it is important that the centering punch be held exactly vertical. This work is best done in the staking tool.

Price, **30c**

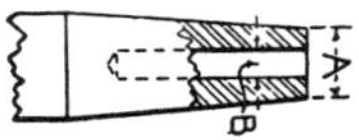

Flat-faced Hollow Punches. Used for the finish staking of wheels on pinions, balances on staffs, end shaking, pressing on hairspring collets, and a great variety of work.

Price, **25c**

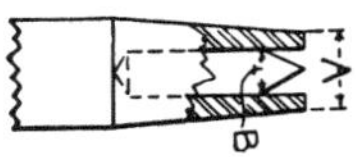

Center Wheel Punches. Used to indent the riveting of safety pinion staffs. If the staff is first properly staked with smooth punches, only four indentations should be made, and they need not be deep. Without these punches it is sometimes very difficult to prevent the staff turning in the wheel. Used in the watch factories.

Price, **30c**

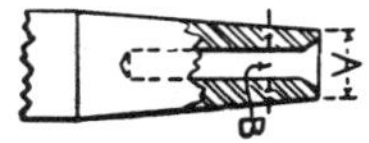

Hollow Taper Mouth Punches. Used for closing holes, hour, minute and second hand sockets. Price, **25c**

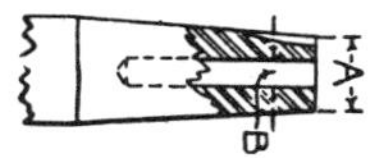

Roller Driving. With hole in center to go over staff, and groove in side to receive roller pin.

Price, **25c**

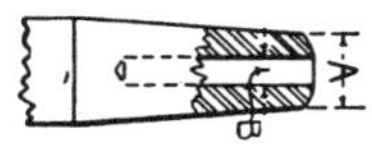

Round-faced Hollow Punches. For staking balances, wheels on pinions, etc. When staking a wheel on a pinion, Swiss style, the wheel should first be carefully pressed down with a flat-faced punch of suitable size. Then a round-faced punch should be used to slightly bend the points of the leaves outward; then a flat-faced punch to complete the riveting. If the wheel is properly fitted to the pinion, there is no danger of bending or injuring the leaves. The faces of our round and flat-faced punches are highly polished.

Price, **25c**

Stumps

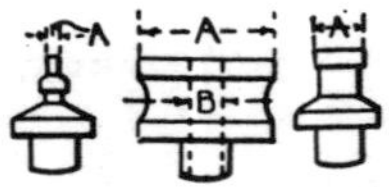

FLAT-FACE SOLID. Used for closing holes in conjunction with round-face solid punches, peening, etc. Price, **20c**

ROUND-FACE SOLID. Used in conjunctin with round or flat-face solid punches for closing holes, etc. Price, **20c**

CROTCH. For supporting cannon pinion while adjusting setting friction. Insert taper brass wire in the pinion; rest side of pinion in crotch and bring down a small round-face punch or the peening punch on the exact place depression is to be made; a light blow on the punch is usually sufficient. If friction is too stiff, lessen it slightly by inserting a round broach in the pinion. Price, **25c**

SMALL CONCAVE. For supporting back end of center arbor, while staking on hands or cannon pinion. Price, **25c**

ROLLER REMOVER. For two-arm balances; has transverse slot to receive balance arm and notch for end of roller pin if it projects. Price, **30c**

ROLLER REMOVER. For three-arm balances; has six radial slots to receive arms and central hole for hub of balance. Price, **30c**

CYLINDER. For supporting lower shell of cylinder while inserting new plug. Price, **30c**

ROLLER DRIVING. Has central hole to receive staff and side groove for roller pin. Price, **30c**

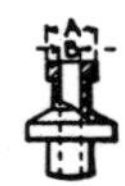

FLAT-FACE HOLLOW. Used for resting wheels for driving out staffs, pinions, etc. Price, **25c**

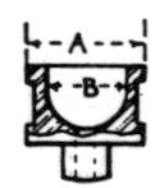

LARGE FLAT-FACE CUP. Used for bumping plates or bridges, end-shaking barrel arbors, etc. Price, **30c**

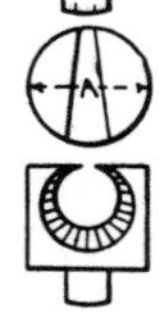

V SLOT. For driving out escape pinions without disturbing wheel bushing, etc. Very practical and useful. Price, **35c**

Frictioning the Cannon Pinion

PROBABLY in no detail of watch repairing, have more plans been proposed than in the matter of frictioning the cannon pinion.

Most American watches are now so made that frictioning the pinion gives but little trouble. Some of the companies have made pinions of many different styles, but we will not need to concern ourselves with all these; consideration of a few typical cases will be sufficient.

First: the class of pinions now very common, in which the tube of the pinion is reduced to produce sufficient resilience of surface. Such a pinion is shown in Fig. 1, in which the surfaces *a*, form the bearing for the hour wheel, and the reduced part *b*, the resilient portion; which may be manipulated in various ways, to friction it on the arbor.

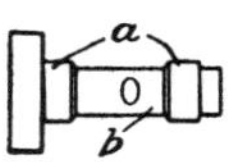

Fig. 1

A very common way is to squeeze it with cutting pliers, producing indentations on opposite sides. This is generally effective enough, but is not neat and workmanlike as we should always try to have our work appear. A pair of pliers might be prepared for this purpose, with the jaws rounded and smooth, which would do the work with the desired nicety.

This work may be done to perfection with the Staking Tool. We place the pinion on a taper brass wire—supplied with all the better class of K. & D. Staking Tools—place the crotch stump in position, and laying the pinion in the notch—the wire serves as a convenient handle—we bring down the peening punch, with its rounded end arranged crosswise the pinion, Fig. 2, and with a light tap of the hammer on the punch, produce an indentation in the pinion as shown at *b*, in Fig. 1. Should we find the friction now too stiff, it may be lessened by inserting a round broach in the pinion, and turning it slightly.

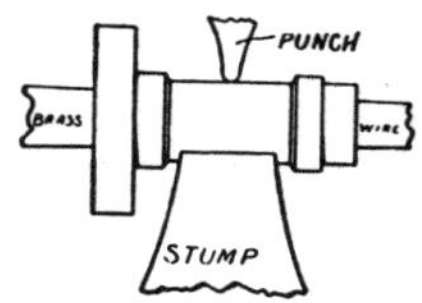

Fig. 2

We must exercise care in this, or the friction may again be too free. We frequently find that the original indentation has been made so far down on the pinion, that it will be loose when placed in position, no matter how much we may indent it. This effect is sometimes produced by the center arbor having too much back taper.

In such cases, we should straighten the original indentation, by inserting a round broach, and then making an indentation higher up. We have had cases where it was necessary to shorten the upper bearing *a*, for the hour wheel, in order to get the indentation high enough to be effective. In a few cases we have found it necessary to modify the back taper of center arbor slightly.

The Elgin pinions having the tongue, may be frictioned by inserting a wire in the pinion, up to about the center of the tongue, and then

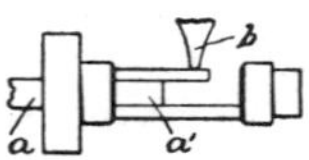

Fig. 3

depressing the point of the tongue with the punch in the staking tool. Fig. 3 shows the conditions; *a* is the wire, the square end at *a*, *b* is the end of the peen punch, resting on the end of the tongue. The pinion is now laid in the crotch stump, when a light tap with the hammer on the punch will slightly bend the end of the tongue downward.

If the increase of friction needed is but slight, the taper brass wire put through, as with the pinion with thin wall, instead of the

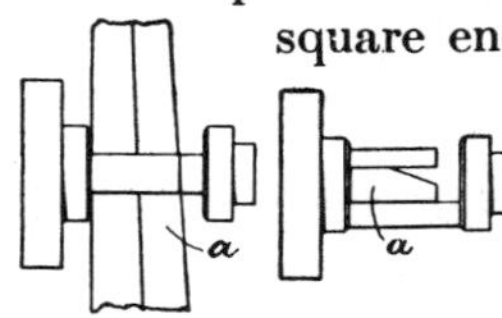

Fig. 4

square ended wire we have recommended, will answer. Another plan which works well, is to make a steel piece shaped in section thus:  and made taper. This may be put through the pinion crosswise under the tongue, as shown in Fig. 4, in which the piece *a* is shown both in plan and elevation. The piece *a* should be hardened and tempered to a blue or spring temper. It will be found convenient to provide the piece *a* with a suitable handle, or what is as well or better, is to file up the form *a* on the end of a piece of steel long enough to serve as a handle. With this tool, the end of the pinion tongue may be bent down, with but little risk of breakage. Attempting to bend the tongue without something under it is pretty sure to result in breakage.

The Elgin pinions with the *transverse* tongues, may be satisfactorily frictioned by inserting a taper brass wire, and bending the tongue down with a punch, used as in the case of the thin walled pinion. In the preceding cases we have used the peening punch; about equally good results may be obtained by the use of a small round-faced punch.

In many stem winding Swiss watches, the center pinions are hollow and the arbor is frictioned in the center pinion. It is generally more difficult to get a smooth and uniform setting friction on these than on the pinions we have been considering, for the reason that the holes in the center pinions are seldom absolutely round, and the walls are thick and unyielding.

Sometimes it is advisable to make the center arbor tight in the center pinion, then make a slight back taper on the front end of the arbor, and friction the cannon pinion as in American watches. But if the arbor is in good condition, and a good fit for the hole in the center pinion, there is an easier way that gives most excellent results.

With a round edge lap, grind a hollow in the center arbor, at about the middle of the part working in the center pinion. Then lay the arbor on a brass stake, or stump in the staking tool, and with the peening punch placed in the hollow as shown in Fig 5, a light tap on the punch *slightly* bends the arbor, with very little danger of breakage. If we have no brass stump, a piece of sheet brass laid on top of a steel stump, answers as well.

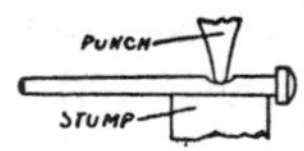

Fig. 5

When the slightly bent arbor is inserted in the center pinion, the walls of the hole bring it straight again. This plan provides a remarkably smooth and durable friction. Instead of grinding, the hollow may be filed, but grinding is much the quicker and better way. Grind the arbor on one side *only*—this is important. Before inserting the arbor, be sure to oil it; setting friction surfaces should *always* be oiled.

Many watchmakers adjust the setting friction much too stiff. It should in no case be so stiff as to reverse the motion of the train, when the hands are set back. The average workman experiences more difficulty, as a rule, in adjusting the friction in watches of the full plate type, for the reason that the walls of the cannon pinion are thick.

A generally satisfactory method is to grind a hollow in one side of the pinion with our round edge lap, as shown in Fig. 6. The hollow should extend nearly through the wall of the pinion. We then slip the pinion on a round taper brass wire, and with the smallest round-faced punch, make a slight indentation in the center of the hollow, resting the pinion during the punching in the crotch stump in the usual way.

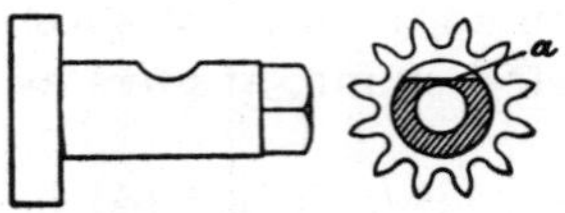

Fig. 6

If the grinding of the hollow is properly done, this makes a good, workmanlike job, and produces a smooth, even friction. In some cases, where the arbor has been damaged by the use of cutting pliers, or other harsh methods, in efforts to friction the pinion, and in which cases it would be difficult to get a smooth, even friction in the usual and legitimate way, it may be allowable to punch the lower end of the hole in the cannon pinion, with the triangular pointed punch used to tighten rollers. The arbor is generally uninjured at the point where the burrs thrown up by the triangular pointed punch would come, and this method is frequently effective, though we don't pretend to like it.

There are some cases—generally very old watches—where the center arbor is so very badly injured, as to make it entirely impracticable to friction by any of the regular methods. When the arbor will admit of reduction sufficient to make it round and smooth and still be large enough for strength, we may proceed as follows: rebore the pinion large enough to receive a steel tube fitting the arbor, and with a wall thick enough to admit of frictioning by indentation in the usual way.

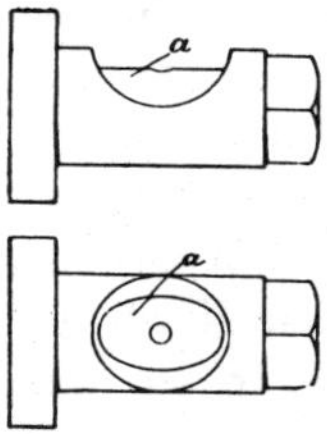

Fig. 7

After the pinion is rebored, and before driving the tube in, we grind through one side of the pipe with our round edge lap; the hollow should extend to the center. We then insert the tube tightly in the pinion, finishing the end flush; then slip the pinion on the taper brass wire and indent in the usual way; Fig. 7, shows the finished job, plan and edge views.

The careful workman notes the condition of the setting friction when he takes the watch down; and if the arbor needs repairing, he does it before cleaning and assembling the watch. If *all* would do likewise, much time would be saved and botch work avoided. Nothing contributes more to *rapidity* and *high standards*, than system in our work.

Closing Holes

IN a general way, the methods of closing holes may be divided into two classes: that by which the metal is upset and condensed, and that by which the metal is gathered in and condensed—as by a taper mouth hollow punch. Each method is best suited to particular cases, for example: with small holes, the first method is best; with large holes, the latter method.

Closing train pivot holes in watches having a low number of jewels, is a job that comes almost daily to the watchmaker; and often many times daily.

There are some workmen who eschew closing pivot holes, and insist on bushing in every instance. Every one has the right, certainly, to decide what course he will pursue; but is it not wiser to exercise discrimination and judgment in *each* case, than to sweepingly denounce, just because a method would be botch-work if followed in *some* cases.

It is a fact that many holes are closed that should have been bushed; so let us try to confine ourselves to the consideration of cases that we may safely call legitimate: holes slightly worn, or originally too large, for example.

Every watchmaker understands that correctly fitting pivot holes are very important if a watch is to run even moderately well. In the first place, holes too large result in *loss* of power, and in addition to this, it makes the power that is transmitted, variable in amount. Therefore, it is a poor policy to put in a stronger mainspring to bring the vibrations of the balance to the required amplitude; we will be building reputation by doing it right in the first place, and generally we save time as well.

Being methodical watchmakers, we observe the nature and extent of the necessary repairs before taking the watch down. When closing of holes is called for, we carefully note the end shakes, and change them if necessary, when we close the hole.

It will be understood that we are considering the closing of holes in watches only. Methods are practiced in the repairing of cheap clocks, which we would not tolerate in watches—even cheap ones. Moreover, the Staking Tool is not generally used in closing holes in clocks; every workman recognizing that less artistic work is required. In practical work, we frequently have to modify the old adage "whatever is worth doing, is worth doing well" to, whatever we do, we will do *well enough*. This means that we will do each job as well as the general quality of the watch requires—we will preserve the harmonious character of the machine.

Having examined our watch, and found that several holes require closing, we proceed as follows: first peg out the holes; this may save imbedding grit or metallic oxide in the metal, which would subsequently cut the pivot; at any rate it's a safe precaution.

Fig. 1 shows in section—through a pivot hole—a part of a watch plate. This shows a pivot hole as it *should be*, not as it often *is*, even in new watches. We usually find a considerable burr at *c*, around the upper end of the hole. This interferes materially with the proper supply of oil to the pivot, and should not be tolerated; let no burr be left on the holes *we* close.

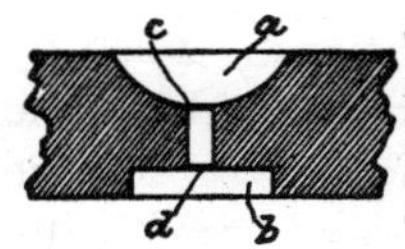

Fig. 1

We select a round faced punch, of slightly smaller radius than the concave sink *a*—this allows it to touch the bottom of the sink *a*, immediately around the hole—see Fig. 2, and a flat-faced stump which will freely enter the recess *b*, of Fig. 1. The depth of the recess *b*, varies with different cases, and frequently there is none at all; but where there is one, it is essential that the stump should enter freely. Placing the stump in the Staking Tool, we hold the plate in position with the recess *b*, on the stump, and bring down the round-faced punch to enter the con-

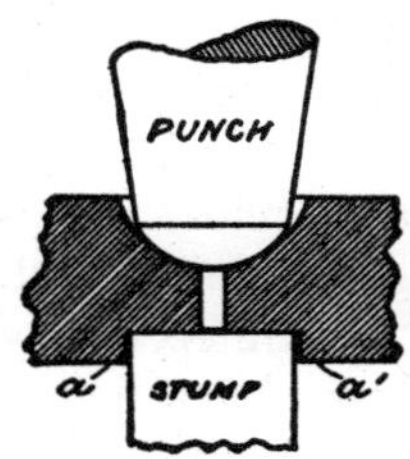

Fig. 2

cave *a*; Fig. 2 shows the conditions. Then by grasping the punch with the fingers, above the top of the tool, and pressing downward we hold the plate in position, and practically horizontal; when a blow on the upper end of the punch with the hammer condenses the metal between the punch and stump; as the hole offers the path of least resistance, most of the displaced metal goes into it.

This process shortens the hole somewhat; but the final result, when properly done, is a bearing at least equal to the original. Our next step is to open the hole to properly fit the pivot; this we do with a round broach, oiled. We should avoid closing the hole in the first instance, beyond what is really necessary to form a good bearing the full length of the pivot; for if we do, we unnecessarily shorten the hole, and make extra labor—both are losses that should be avoided.

The action of the round broach in opening the hole invariably throws up a burr at the ends of the hole, which should be removed. To remove it we need a special countersink or chamfering tool. Such a tool is shown in Fig. 3, and is made of a piece of steel wire about 1.5 mm. in diameter, with the end turned to about

Fig. 3

the angle shown, when it is hardened and tempered to a straw color; the pointed end should be polished; then we grind away almost half the diameter, as shown by dotted line; this surface need not be polished, though it should be ground *smooth* on a fine Arkansas or Jasper stone. It should be fitted to a light taper pegwood handle, although if one wishes he may have something nicer than pegwood for a handle, such as ebony or rosewood, or aluminum, but a good round piece of pegwood, tapered as shown, leaves nothing to be desired in the way of convenience. An hexagonal piece applied as shown at *a* and *b*, the latter being an end view, makes the tool easier to pick up, and prevents rolling.

After we have opened the hole with the round broach, a light application of the chamfering tool we have just described, will remove the burr, leaving the ends of the hole in perfect condition; if we choose, we may lightly insert and turn the round broach again after using the chamfering tool, but it will rarely be necessary.

The broaches one is generally able to buy in the market are not very satisfactory; it will pay us to fix them up before attempting to use them. We select a pentagonal or cutting broach of the required size, and fixing it firmly in a light pin vise, or broach holder, we lay it on a flat piece of cork or fungus, or a piece of soft pine, held in a vise, and with an India oil stone slip, grind the five sides flat and bring the angles up nice and sharp. The finishing strokes should be lengthwise the broach, which gives it much smoother cutting qualities.

Then we select a "round" broach, so-called, for they are never really round, as nearly perfect as possible, and grasping the shank in a lathe, we grind it practically true, in taper and roundness. This is easily done by holding a flat piece of cork under the broach as it revolves, and the oilstone slip is held over it; not pressed closely down, but allowing the oilstone to rest upon the cork back of the broach, the two forming an acute angle; the space is then narrowed, until the stone just touches the high spots on the broach as it revolves; this process soon brings it practically true.

This leaves the lines running *around* the broach; we want them to run lengthwise; we do this by rubbing the broach lengthwise, while held on a smooth piece of soft wood, with an emery buff of suitable grade. By using a rather coarse emery buff a considerable degree of cut can be given to a round broach, when used in soft metals like brass or German silver.

Our round and cutting broaches should be fitted with a light, straight wood handle, such as a piece of pegwood.

If we have a suitable box to contain our pivot hole tools, and label it so we will recognize it easily, and always keep the tools in it when not in use, we will find it a great saving of time. The tools are always together, and we are able at all times to put our hand right upon them; *system* is a great thing for any workman to tie to; he turns off his work rapidly, saves lots of nervous energy, and often economizes greatly on the swear words.

The hole, as we have closed and finished it, will be about as near perfection as a metal hole can be; the slight chamfers we have made at the ends serve to hold the oil around the pivot, even to the last particle. In opening the hole with the round broach, we must take pains to maintain the broach upright; if we fail to do this, the pivot may pinch in the hole when we set the watch up.

In case we find on our preliminary examination that the end shake should be lessened, we should close the hole just as we did in the first instance; then using a large flat-faced stump to support the plate on the surface *a*, *a′*, Fig. 2, with a blow on the punch, we drive the metal downward. It is generally wise to use the small flat-faced stump again after this, to harden the metal and flatten the surface around the hole. The hole is then opened and finished in the usual way.

If the punch used in closing is highly polished, the oil sink around the hole will be smooth, and hardly need any polishing; it will be more than equal to the original hole, as they are left in most American watches.

The business end of all regular K. & D. punches are highly polished—by "regular" is meant all the highest grade tools, as distinguished from the "Specials".

In case it is required to polish the oil sink after closing, it is easily done by means of a soft pine stick and rouge, or almost any polishing powder.

To return again to the matter of punches; it will be found that all round-faced punches are less rounded than we have recommended; it will be worth while for the watchmaker to re-shape and re-finish all his smaller round-face punches, up to about No. 40 Stubs' gage, to conform to the rounding shown in Fig. 2. Henceforth, all the round-faced K. & D. punches of these sizes will be made in this way.

Before leaving the subject of closing pivot holes, we will describe a device for the ambitious workman, who has time, or finds time, to make a few special tools, that save on the job many times over what it costs to make them.

In the opening of holes as usually practiced, considerable time is spent in finding suitable broaches, in the first place, then in the repeated trials of the pivot in the hole.

The device here shown enables us to open the hole the first time to the correct size. We first prepare a round broach, as we have already described, grinding it to as true a taper as possible; we finish it by lengthwise grinding, as though we were to use it as a round broach.

We now convert it into a cutting broach, by grinding away slightly less than half the diameter; about 45–100 of the diameter will be found to give good results. This flattened side should also be ground to leave the lines running lengthwise, and all burr carefully removed. This broach will produce very round, smooth holes.

Our next step is to provide a suitable stop, which may be set at any point on the broach; the arrangement is shown in Fig 4. It consists of a steel, brass, or German silver rod *a*, of about 2 mm. diameter, fitted to a neat, light taper handle, *b;* the exposed part of *a* should be equal in length to our broach, or slightly more. The broach is fixed securely in the outer end of *a;* the whole thing should be true when rotated.

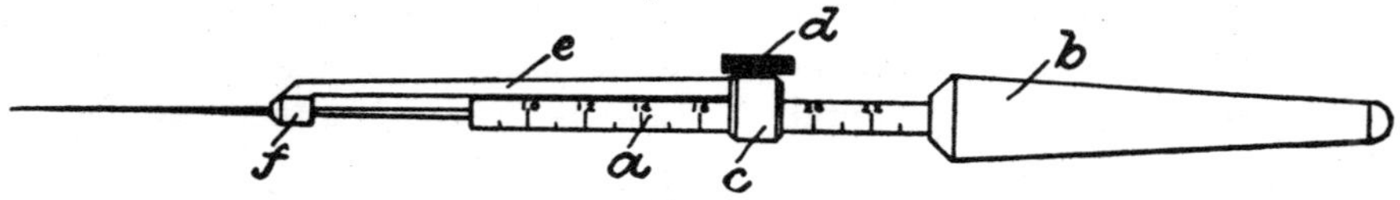

Fig. 4

We now have the part *c*, a neat sliding fit on *a*, capable of being tightened at any point by the knurled screw *d*. From the part *c*, an arm *e* projects, carrying at its outer end a light steel tube *f*, having an inner diameter slightly greater than the broach at its largest part. When the part *c*, is close against the handle, the inner end of the steel tube *f*, should just clear the end of *a*.

For our own part, we prefer *e*, and *f*, to be of tempered steel; *c*, *e*, and *f*, being fastened together with hard solder, and afterward hardened and tempered, then ground to our liking. Of course each workman may follow his own methods, in this respect; some will use soft solder, and "get there just the same". Whatever method we adopt, let us give the device that finished look, as though made by machinery, and highly skilled workmen, and not painfully gnawed out by mice.

Now to make use of the device; we measure the pivot, then adjust the end of *f* to a point on the broach, measuring the same as the pivot. Now if the hole has been closed, all we do is to run the broach in until stopped by *f* when the hole should be of the same diameter as the pivot. To have the hole as nearly straight as possible, the broach should be inserted from both sides.

Now the final burnishing by a suitable round broach should give the required freedom. What is still better is to have a *round* broach also, fitted up in the manner described; for in many cases of hole closing the round broach *only* should be used to open.

It is obvious that the rod *a*, may be graduated to correspond with our gage; then upon measuring the pivot, the broaching device may be instantly set to the required size.

It will be seen that this device is nothing more than a slight variation on the taper needle jewel gage; but that does not detract from its practical value. Another advantage of having a round broach fitted up like this, is besides a great saving of time on ordinary opening jobs, is the fact that we may give a known amount of freedom to any pivot, thus eliminating the element of "guess work" entirely.

A neater design of the micrometer broach would be to make *a* of steel, and arrange the broach to slide within it; in this way we would utilize the end of *a* as the stop, to limit the distance to which the broach might be inserted. Such construction would be more difficult, however, and we would advise the workman to try one as shown, first; then make any improvement he sees fit.

Closing Larger Holes.

The closing of holes in main spring barrels and the like is a different problem. Some workmen close, or attempt to close, such holes with round faced punches; the result is never satisfactory. Fig. 5 shows in section the arbor hole of an ordinary main spring barrel. The boss which surrounds the hole, rising above the level of the bottom of the barrel, varies considerably in shape and thickness in different barrels, but in any case there is nothing to prevent making the boss of the shape shown in Fig. 5, which may be very quickly done with a hand graver while the barrel is held in a wheel chuck in the lathe. We then select a hollow taper mouth punch which will go over the boss, and down to about the point shown in Fig. 6, and with the barrel on the large flat faced stump, a good blow on the punch closes in the boss, and also the hole, not only at the small end of boss, but throughout the length of the hole.

Fig. 5

Fig. 6

In case we have to reduce the boss, it is well to make it of a taper to conform to our taper mouth punch; the nearer the taper of the boss conforms to the taper of the punch, the more uniform will be the closing of the hole, throughout its length.

Now if the wear was only *slight*, the closed hole may be opened up with a round broach; the burrs nicely removed, and the barrel should be found to run practically true, and satisfactory. But if the hole is much worn, the barrel should be chucked in the lathe, and the closed hole trued up; it is preferable to true the hole in the cover last. While we have been speaking of closing the hole in the barrel only, of course the same applies to the cover.

There's a reason why the barrel should be trued in the lathe if the hole is much worn; mainspring barrel holes never wear central, for the reason that the pressure on the arbor varies greatly with the position of the outer end of the mainspring, as the barrel revolves. When the end of the spring is directly between the center pinion and the barrel arbor, the pressure is but a small part of the pull of the mainspring; but when the arbor is directly between the end of the spring and the pinion, in other words, when the barrel has made half a revolution from its first named position, the pressure on the arbor is about twice the pull of the spring. This causes a very unequal wearing of the hole.

In some American watches, which were so poorly designed as to require an abnormally stiff spring, the cover boss is very short; in many cases so short as to make it difficult, if not impossible, to close the hole as we have described; in such cases we may either bush the hole, or resort to the round-faced punch, which still further shortens the already too short hole.

The better way would be to shorten the hub of the barrel arbor, to allow a cover bearing of sufficient length, and bush the hole to conform. It is never necessary to have the width of the arbor hub for the spring to wind on more than one third the width of the spring. Often in American watches we see the arbor hub chamfered off on each side, making the face no more than one third the width of the mainspring, yet with bearings very thin; the chamfering is an absolute waste of space which should have been devoted to longer barrel bearings.

In opening comparatively large holes, like those of a barrel, it will be found of advantage to insert a narrow strip of paper beside the round broach; it produces very much smoother work than when done without, besides the hole will be straight; this is because the paper yields, allowing the broach to lie closely to the wall of the hole; the

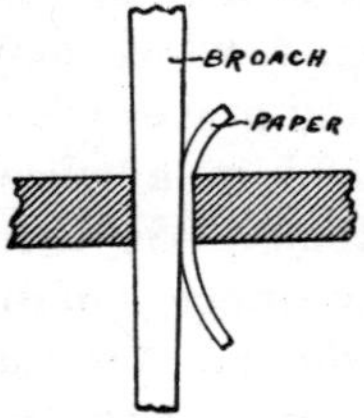

Fig. 7

conditions are shown in Fig. 7, where the taper of the broach is much exaggerated for clearness.

A narrow shaving of soft wood answers as well as paper; in fact, any substance of a character similar to wood or paper may be used. The same idea may be made use of to open a hole straight with a taper cutting broach. If the hole is straight to begin with, it may easily be opened, and remain straight; the tendency is to preserve the original form of the hole, whether straight or taper.

CLOSING HAND SOCKETS.

The sockets of hour hands may usually be closed within certain limits, in the same manner we closed the barrel hole; shown in Fig. 6. There will be this difference, however; the hand socket will be closed at the end only, and it should, therefore, be closed somewhat more than desired, and then opened up by filing, grinding, or turning, to properly fit the pipe of the hour wheel. Minute hand sockets are closed in various ways, depending largely on conditions. If the hand has a brass bushing, projecting considerably on one or both sides, it may be closed with a taper mouth hollow punch. If it is an all steel socket, or does not project at all, the round face punch serves best. Second hand pipes are sometimes closed with a small taper mouth punch, or by forcing the socket into one of the chamfer mouth holes in the staking tool die; this closes the socket at the end only, which is likely to produce a wabbly fit. A generally more satisfactory way is to insert a cutting broach in the socket, and indent its side, just as we do in frictioning a cannon pinion; if the closing needed is but slight, better insert a round broach instead of a cutting broach.

To perform this operation, we will need to lay the socket on the crotch stump, to one side of the notch; or we may use a flat-face stump. The reason we cannot use the notch is, that it would be too deep for most sockets. We have sometimes used the notch, and a small round-face punch to do the indenting; the result is satisfactory.

But for tightening a second hand, an old pair of cutting pliers with the jaws ground rounding, is much more quickly used than the staking tool, and the result is quite as good. Some watchmakers close second hand sockets by pinching them in a wire chuck. This method closes them very nicely, if one cares to use his chucks for such a purpose.

Bushing

THE Staking Tool is also a great convenience for riveting bushings, etc., for by its use we are enabled to rivet the bushing very closely flush with the plate or bridge, with no danger of marring it; this allows us to retain the most hardened part of the bushing for the purpose of a bearing.

Most young watchmakers make the mistake of trying to rivet a bushing securely into a round hole; such a bushing is liable to turn while being bored and finished, and waste all our labor.

The hole in which a bushing is riveted should be provided with grooves lengthwise, into which the metal of the bushing is forced when riveted; shown at *a*, *a*, Fig. 1; these grooves are conveniently formed by forcing a cutting broach into the hole, and rocking it slightly from side to side—do not turn it. This should be done from both sides of the plate. You will find examples of this in old English watches, both in case of the bushings and the pillars.

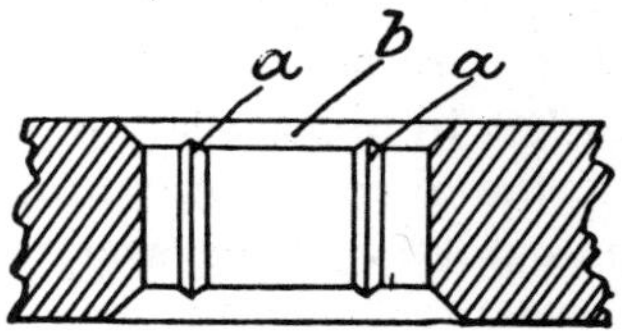

Fig. 1

In putting in a bushing, it is better to use brass, not too hard, and make the bushing rather long, to be shortened by the riveting process; this allows the metal to flow freely into the previously formed grooves, and the riveting condenses and hardens the metal sufficiently. When we speak of the metal flowing freely, we do not mean, of course, that the metal is melted; flow is a term used to express the movement, among themselves, of particles composing a body, when under pressure, hammering, etc.

Suppose we wish to bush, say a center hole. We first broach out the hole to a size sufficient to give a safe thickness of wall when the hole is opened. No hard and fast rules can be laid down for this; the workman must rely upon his own judgment; a small hole requiring a much thicker wall, proportionately, than a large one. The hole should be chamfered rather deeply, as shown in Fig. 1, at *b*; if these chamfers are made shallow, as, for example, with a wheel countersink, the end of the bushing is likely to flake up after finishing; the deep chamfer also gives a much more secure riveting.

The plug should be turned to fit the hole tightly; the ends cupped as shown in section in Fig. 2; and of a length sufficient to allow for riveting, which can only be learned by experience. Now with the bushing resting on a flat-faced stump in the Staking Tool, we bring down a round-faced punch and with a blow of the hammer spread the end of the bushing; we then reverse the plate and do the same on the other end of the bushing. After this we finish with a flat-faced punch. The job is then mounted in the lathe perfectly central, and the hole opened to fit the pivot or bearing, the end shake adjusted, etc. The order of these operations is to fit the pivot first, then adjust end shake, lastly stripping out the oil sink, and lightly chamfering the end of the hole next the shoulder of pivot or bearing.

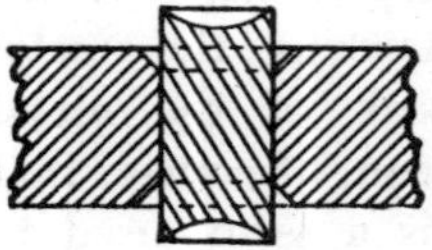

Fig. 2

Many watchmakers have trouble in stripping oil sinks, jewel settings, etc., satisfactorily. Of course a sapphire stripper is the best tool for this purpose, but very respectable work may be done with a steel graver, if we sharpen it properly. We need a graver of good quality, ground to the angle we desire, and finished with perfectly straight edges, *i. e.*, unless we require a curved edge. We then finish the end of the graver on a jasper stone, such as used for polishing jewel settings on the flat face. We do not need to finish the whole flat face of the graver; only a narrow strip along the edge on *both sides;* this finishing is done by rubbing the graver *lengthwise the edge;* thus the microscopic lines left by the jasper stone run at right angles to the direction of cutting; the result is a brilliant cut left by the graver on soft metals such as brass, gold, German silver, etc. This graver should not be used except to remove the final chip or two, or its brilliant cutting quality will soon be impaired.

Here is another case where diamond powder is most useful to the watchmaker; keeping the jasper stone in perfect condition; for they get dull after a while, and need resharpening. Other material, such as carborundum, may be used with very good results, but it does not give the quality to the surface that diamond powder does. The reason will be plain if we examine the shape of carborundum and diamond crystals under a magnifier.

Then there's a class of bushings which are best fastened by means of soft solder; such, for example, as bushing the recess in the pillar plate of a Howard watch, in which the barrel has its bearing. Sometimes these barrels are not too smooth, and many watchmakers forget to oil them, with the result that sometimes the wall of the recess is very badly worn. Of course, such a bushing could be fastened by rivets or screws; but we think this a case where soft solder serves admirably if properly used. Three things should be very carefully observed: first, be careful not to overheat the plate—use easy flowing solder; second, thoroughly wash the plate after soldering, and after dipping in the cyanide solution; third, do not leave a *particle* of solder visible on the job, after it's finished. To have solder flow readily and thoroughly, it is important to have the work *clean*, and to see that the fluid has gone to every part where we expect the solder to flow.

In nearly all cases where bushing is required, the pivot or bearing needs refinishing also. It may not be amiss to consider briefly the refinishing. If the pivot or bearing is actually cut enough to destroy its cylindrical form, it is nearly always best to *turn* it true, before attempting to grind and polish.

In this day every watchmaker should have a pivot polisher; they are great time savers. Let us suppose it is a barrel we have to bush, and that the arbor is considerably cut and worn. We mount the arbor in the lathe to run truly, and with a sharp graver turn the bearing true, and take out nearly all the damaged part, leaving only enough to allow for grinding; thus avoiding unnecessary reduction of the bearing.

Properly filing the grinding lap is very important, before trying to use it on a pivot or bearing. Particularly if we have not turned a perfect corner at the root of the pivot or bearing, the lap will then need to be in perfect condition.

Before filing the lap it is turned true and of the exact form we require. The filing, be it understood, is to enable the lap to hold the grinding or polishing powder more readily, also to produce smoother work; for we shall see upon reflection that a lap just as left by the turning tool would have slight grooves in it, running in the direction of motion of the lap when in use, and these grooves would be reproduced in the work; or to be more exact, their reverse would be produced.

Filing a lap, it is true, has a tendency to destroy its truth; but at the same time it enables it to do much better work, and if properly done it is surprising how many times a lap may be filed, and yet run practically true.

For filing laps a No. 8 flat file, about five inches long, may be used. It should be fitted with a proper handle, which should be straight with the file. A very good plan in fitting handles to files is to heat the tang of the file red hot, and burn out the handle to a correct fit. Of course, in heating the tang of the file, care should be taken to prevent the temper of the cutting part of the file being drawn.

Many watchmakers seem to prefer using files without handles; this may do for much of the work that comes to hand, such as pointing pintongs, etc., though even here we should think comfort would suggest a handle. Speaking of comfort reminds us of a "self-made" watchmaker we once knew. He used his files without handles, presumably because they came that way. We observed that in using a flat file, he grasped it by the end opposite that intended for the handle, thus pushing it backward. When questioned as to his reason for so holding it, replied: "I guess I don't want that sharp end to stick in my hand".

Undoubtedly in much of the work for which

small files are used, they are fully as convenient without handles; but flat files of the dimensions used on laps, in fact all files of the larger sizes, should be fitted with proper handles.

This is one of those points, seemingly unimportant, even considered by some a matter of fancy; but we can say decidedly that such is not the case. The young workman should be particular to acquire *right methods;* such is the basis of good work, and although a file handle is a small thing, it contributes a great deal in turning out a good job, when accurate shapes and surfaces are required.

To come back to filing our grinding lap. We first file the flat face *a*, of the lap, Fig. 3, holding the lap with the fingers and thumb of the left hand. The file should be brought flat against the lap with slight pressure, and moved in the cutting direction only an inch or so, when the lap should be turned partly around in the fingers, and another stroke made; this crosses the lines somewhat and tends to keep the lap true.

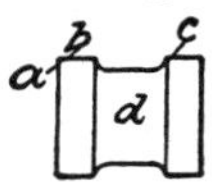

Fig. 3

We next file the surfaces *b*, *c;* this is a little more difficult to do really well, but if we go about it right, and exercise *care*, we will get a satisfactory result. The lap must be held in the fingers; for by this means it is maintained closely in contact with the file through the entire stroke, and the hands instinctively cooperate, producing much truer work than could possibly be done with the lap held solidly.

In filing the side of a lap, we have always been told to file it lengthwise; but we shall make a slight variation on this procedure, and to our mind, produce better results. Besides, to file a lap lengthwise, we produce more or less flattened surfaces, which destroy the truth and smooth working quality very quickly.

We will file our lap in the direction of a helix, or as though we were following a screw thread of rapid pitch. To make this still plainer, we show the arrow in Fig. 4, which indicates the direction in which the file should be applied and moved.

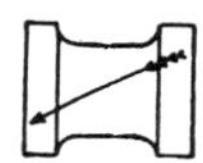

Fig. 4

Another very important point is: we should

always file *toward* the face *a*, of the lap which we have already cross filed. You will notice when you follow directions, that a slight burr is turned over the face of the lap by the helical filing of the side surfaces; this burr serves a useful purpose; leave it there.

In doing the helical filing, we should hold the lap with the fingers and thumb of the left hand, with as much as possible of the lap projecting above the fingers, when we apply the file; taking care to maintain it constantly in contact with both surfaces *b*, *c*, and make the stroke with the file in the direction of the arrow. The lap should then be shifted through a slight angle, and the operation repeated.

The knack is soon acquired, and then a few strokes of the file serves to put a lap in good condition. Of course we need to true them up occasionally, for the most skilful filing will in time get a lap out of true. The workman should, by all means, have a special chuck with solid nose for turning and truing laps on. This is merely a solid chuck, which should be a good fit for the lathe spindle, and having a taper end projecting, which is exactly the same as the end of our pivot polisher spindle where the laps fit.

With such a chuck, and suitable reamers to ream the lap blanks, it is an easy matter to make our own laps; every watchmaker should do so, for the tool makers charge a rather high price for them; or, at any rate, with suitable appliances one can save money making them for himself. A good way is to have a dozen or so of blanks, of various materials, drilled and cut off at some machine shop; these will cost us but a few cents each, and save us a great deal of work, as all we have to do to make a new lap is ream the hole, place the blank on our lap truing chuck, and turn it to any form we desire. It is hardly necessary to say that in making or truing laps, a slide rest is very desirable.

Now let us suppose that in turning off the bearing or pivot there was a slight rounding or fillet left in the angle of the shoulder, which we could not quite remove—although we must be able to do it when we are really skilful with the graver—we desire, of course, to have the finished job as near perfection as possible; so perfect that no rounding can be seen in the corner, even with a strong glass.

Let us suppose we have our cast-iron lap filed up properly; we will grind that square-shouldered pivot, and try to remove perfectly the slight rounding in the corner at the root of the pivot. We place the polisher in position, adjusting the lap to stand a short distance, say one millimeter, away from the pivot. We see that the belt is adjusted to run properly; and it should be so adjusted as to draw the lap away, rather than *toward* it. The belt should not be too tight; tightness sets up unpleasant vibrations in the spindle, which are also detrimental to the work. It is best to do all necessary adjusting of spindle, belts, etc., before applying the grinding material.

This being done, we apply fine oilstone powder to the lap by means of our steel spatula. The powder may be mixed to a paste with oil; and for the job now in hand it is best not to be too thick. A little experience sets us right in this matter. We are now at the point where we may either spoil our pivot, or do the work in just the right way. The powder paste should be spread thinly and evenly on the lap; we now bring the lap in contact with the *shoulder* of the pivot, then with the pivot itself, the lathe running *slowly*.

But slight pressure should be put on the lap, and in this case, where we have a rounded corner of the pivot to square up, no traversing of the lap along the pivot should be done. It is here that the slight burr left on the face of the lap by filing gets in its useful work. If we do not bear too heavily on the lap, we shall find on inspection of the pivot that the rounded corner has been entirely removed; that the pivot is straight and the shoulder square. As in this job we are simply concerned in producing a perfect pivot of indefinite size, we will now proceed with the polishing.

One skilled in the use of the pivot polisher can put a fine polish on this pivot with a lap made of almost any soft metal, or bell-metal. But to succeed, a grinding lap, we think, for polishing, a block-tin or zinc lap about the best. Block-tin laps are very tender, and for the beginner we would recommend the zinc. The watchmaker may be unable to buy one, in which case he may make his own. Heavy sheet zinc, such as used for battery plates, is suitable; we must be careful to get *rolled* sheet zinc and not the cast plates; the latter are generally amalgamated with mercury.

A convenient way to make such a lap is to make a brass tube fitting the taper of our pivot polisher spindle, and soft solder it into the zinc disk, which is generally one quarter-inch thick. We then turn the lap into shape; the diameter of such a lap may be somewhat larger than the cast-iron lap if one chooses. Of course, such laps will be too short to admit of holding in the fingers to file the side, although it may be so held to cross file the face.

We get over this difficulty in this way: put the lap on the truing chuck; now adjust the T rest across the lathe bed at a distance of from two to three inches from the face of the lap, and at such height that when a flat file is laid on the T rest, with its end resting on top of the lap, it comes parallel with the lap. Now, by running the lathe very slowly, and slowly moving the file back and forth, with a gentle pressure, the side, or edge, of the lap is soon cross filed very nicely. If one prefers, a roller may be provided to support the file instead of the T rest; but the T serves very well indeed. As it acts merely as a guide for the file, very little, if any, is removed from the T by the file. A roller is convenient, however, and as this is a very practical way to make such laps, it will pay the workman to provide himself with one; Fig. 5.

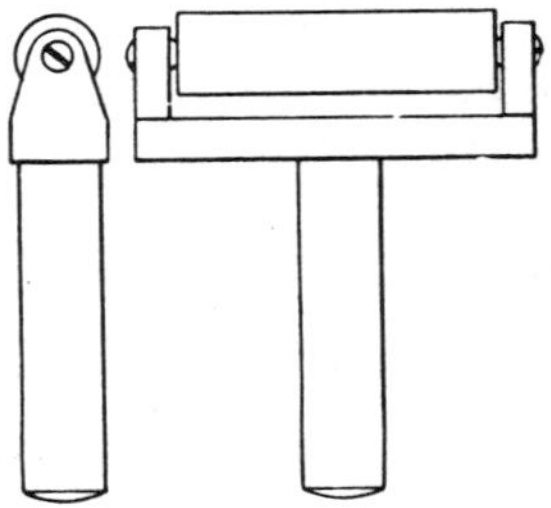

Fig. 5

Assuming that our polishing lap is complete, we will polish the pivot. The same rules in regard to adjustment of lap and belts should be observed, as in the case of the grinding lap. The pivot having been perfectly cleaned of oilstone powder, by pith, and finally "watchmakers' putty", we now apply a very thin coating of sapphirine, mixed with oil, to the face and sides of the lap, by means of our steel spatula, and bring the lap up against the pivot, running the lathe at moderate speed. The lap may be traversed along the pivot slightly, but nothing is gained by trying to traverse the lap on the shoulder.

If the right quantity of polishing material has been applied to the lap, but a very short time is required to bring up a deep black polish on the pivot and shoulder. The surface of the lap and pivot will turn black before the perfect polish appears. If the grinding operation was properly done, our pivot will be straight, smooth and perfectly polished; there will be no visible rounding in the corner at the root of the pivot; the shoulder will be perfectly flat and brilliant; in short, the pivot will look like a perfectly polished cylinder standing in the center of a circular mirror. In all grinding or polishing work on pivots or staffs the work and lap should turn in the *same direction;* the points of contact will then move in opposite directions, insuring the best work.

As a rule, only *moderate speed* should be used; and but light pressure of the lap against the work.

It seems hardly necessary to say that in cases where the bearing must be refinished, it should be done before the bushing is opened up and finished. See article on Closing Holes, for smoothly finishing large holes with the round broach.

Stretching or Swageing

THERE are many operations in watch work which come under this head, for example: lengthening the horns of a fork, or lengthening the entire fork.

Watchmakers should study the theory of escapements; be able to design an escapement, correct in all its actions—not a mere copy of some other drawing. The workman with this degree of knowledge, after examining an escapement, knows with certainty what changes are required to put it in good order, or the best order practicable; for some escapements are so poorly designed as to make it impossible to make them really good without rebuilding.

Suppose we have a Swiss watch, in which we find the notch of the fork much too wide for the jewel pin, and the pin already filling the hole in the roller, and as wide as it should be for favorable action. The obvious procedure is to close the notch in the fork, which may be done with perfect safety, if we proceed in the right way.

The type of fork in which the fault mentioned will most frequently be found, is the kind without a guard pin; the guard function being performed by an abutment on the lever, Fig. 1; the horns, *b*, of the fork being below the plane of the fork body, *c*.

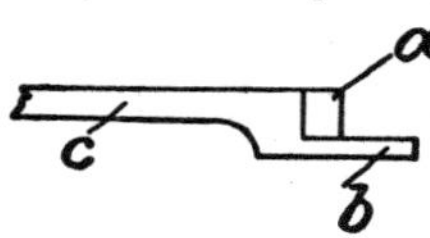

Fig. 1

We should first draw the temper of the fork by grasping it with a heavy pair of tweezers near the fork end, to prevent drawing the temper any further back than necessary, the horns being covered by a folded strip of copper or brass, thin and small; it is well before heating the fork to smear it with castile soap to prevent oxidation as far as possible, or convenient. The heating may be conveniently done in the flame of a small lamp; it should be in a darkened place—*i. e.*, not in full daylight—and the fork heated to a dull red—a red just visible—and allowed to cool slowly. It should now be very soft; but before attempting to close the notch, we should test it for hardness, by means of a knife, graver, or file.

Having ascertained that it is really soft, we place it on the crotch stump to one side of the notch, and with the swageing or peening punch brought down on top of the horn, as shown in Fig. 2, a light tap on the punch bends the horn in slightly; the fork is then reversed and the operation repeated on the other horn. This, of course, makes the notch narrowest at the mouth, so the closing must be enough greater than required to fit the pin, that it may be ground out parallel. (See the article on the Use of Micrometer Stop, also description of Stop, page 66, catalog section.)

Fig. 2

If the horns were at first just the right length, they will now be too long, but with the right appliances—which are very simple—and the necessary knowledge—which is easily acquired—they can be quickly corrected.

About the most convenient thing for grinding out the notch is a thin edged, soft steel or copper lap, charged on its flat face with diamond powder. Such a lap is shown in Fig. 3; it is $1\frac{1}{4}$ inches in diameter, and $\frac{1}{8}$ inch thick; *a* is the charged face. Such laps are preferably mounted on tapers; a sectional view is shown in Fig. 4, which shows that the central portion of the face is turned out below the level of the remainder; this is because only a comparatively narrow ring near the outer edge is used, and such shape results in a saving of diamond powder in charging. The dimensions given are not imperative; they are taken from a lap which we have used for years with satisfactory results.

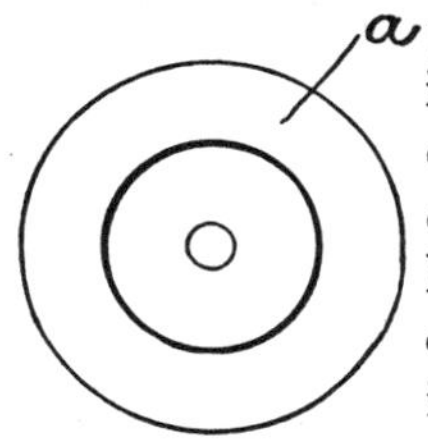

Fig. 3

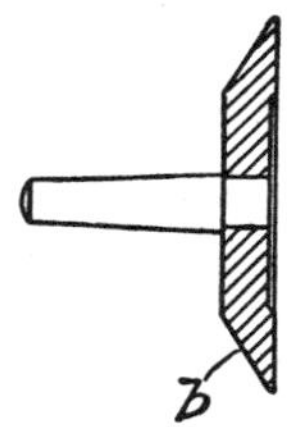

Fig. 4

In making such a lap, it is better to do the first thorough charging, before turning the chamfer, *b*, at the back; this makes the lap more rigid and better able to withstand the pressure of charging.

The most convenient, quickest, and most economical method of charging such laps, is by means of a small hardened steel roller, with rounded face, shown in plan in Fig. 5. The pin, *a*, upon which the roller turns should be hardened and tempered light straw; the roller turns on the pin, the pin being fast in the frame. It may be held fast by being a close fit in the frame, or by a set screw.

Fig. 5

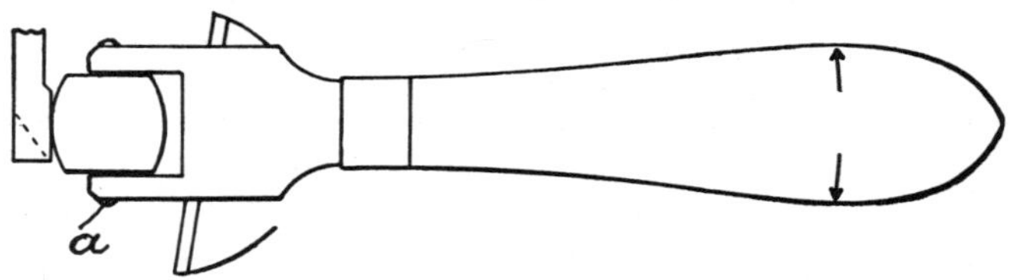

Fig. 6

The holder is fitted to a substantial handle, and in use is held in contact with the lap at about midway its height, the frame of the roller being held on top of the T rest, and the handle carried slowly through an arc of a circle, as shown by arrows, bringing every part of the roller's curved face successively in contact with the lap; the conditions are shown in Fig. 6.

In making these laps, the face, before charging, should be turned as true and smooth as practicable, and then ground smooth and flat on a ground glass lap, with oilstone powder mixed with oil. Such a lap is the finest thing imaginable for grinding small drills, shortening roller pins, pallet stones, etc. When we say shortening pallet stones, of course we mean grinding off the unfinished end; although with proper appliances for holding the pallet stones, and polishing laps for finishing, these same soft steel or copper laps are used to grind the angles on pallet stones, etc. To treat the subject here, however, would lead us too far away from the subject of our discourse.

Before attempting to grind out the notch of the fork, the horns must first be rehardened and tempered. We should have said, that before softening the fork in the first place, the pallets should be removed, if they are separable from the fork, which they generally are. To harden we hold the fork in the same way we did in annealing it; have the fork well smeared with castile soap, and having the alcohol lamp in a darkened place well protected from drafts—for such work cannot be safely done in a flickering and flaring flame—we heat the horns up to a dark cherry red—not yellow or orange—and cool it in water, not too cold; water at the temperature of the ordinary living room, or about 70° is best. The water must be held very close to the flame; a small test tube, or bottle similar in form, is convenient to contain the water.

It is admitted that in softening and rehardening a part only of a fork, that there is left a small part which is soft; but this is of no practical importance; the soft place comes in a part of the fork where no wear occurs, and where it is large enough to withstand all legitimate service without bending.

We now remove any slight discoloration, from the heated part of the fork, by means of fine diamond powder, applied on a flattened piece of pegwood; a tiny piece, the size of a pin point being sufficient. If we have no diamond powder, we scour the discoloration off with fine oilstone powder mixed with oil; restoring the polish, after the notch and horns are reshaped, and refinished, in a manner which we shall describe in our article on polishing. It will be understood that the original finish of the lever is not injured except for a small distance at the fork end.

To come back to grinding out the notch of the fork we have closed. The fork may be

Fig. 7

held by the pallet staff, by means of a K. & D. second hand holder No. 39, or what is still better, a special boxwood clamp shown in

Fig. 7. By means of this clamp we may grasp the fork flatwise at any part of it without marring the polish; although it is always well enough to put a piece of tissue paper around the fork, before closing the jaws upon it.

Keep the face of the diamond lap wet when in use; hold the fork slightly *below* the center of the lap, rather than above; in other words, a line drawn through the fork body, if continued, should pass below the center of the lap. This makes the grinding action disengaging instead of engaging, and the fork will not be forced into the edge of the lap, to the possible injury of both. The conditions are shown in Fig. 8. No rest is needed to support the clamp, other than that afforded by the hands as they rest upon the bench, or upon a block or small box, if the lathe is very high.

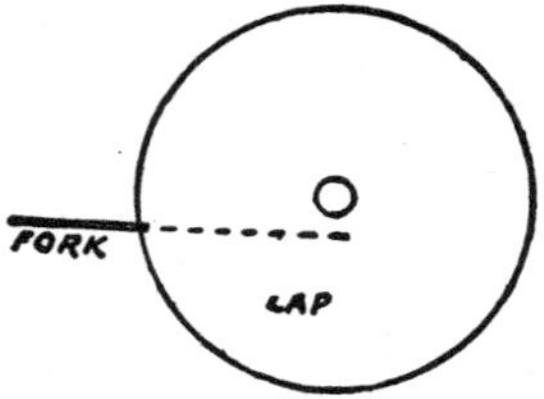

Fig. 8

In grinding out the notch, care should be taken to avoid overdoing it, for diamond laps are quick cutters. The sides of the notch should be slightly rounded as shown in Fig. 9, at *a*, *a′*; this is an end view looking directly into the notch; this form allows a slight lack of uprightness in the roller pin or other parts, without binding. The polishing is done with a similar lap made of boxwood, upon which diamantine, sapphirine, or other polishing powder is used, mixed with oil. Still better is very fine diamond powder; and a better material than boxwood for the lap, is tortoise shell.

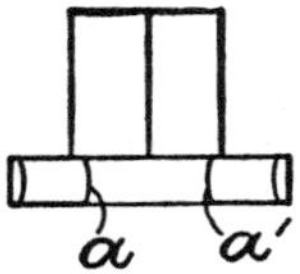

Fig. 9

In another article we shall give directions for grading diamond powder, for the benefit of those unfamiliar with it This material is so very useful to the watchmaker, so rapid in its action, that every workman who aims to be skilful in his art should be familiar with its use, and equipped to do the simpler operations, at least. Opinions differ as to the shake that should be allowed between the roller pin and the notch of fork. We believe the factories allow about .03 mm. or slightly more than .001 of an inch. Some watches in which the pivots fit the jewels poorly, more shake would be required.

We next grind the horns to correct length. We frequently find in this class of escapements, the horns very wide, and ground curved inside; this is quite unnecessary; in single roller escapements, even when poorly designed, the horn is responsible for the safety action only through a very small angle, compared with the double roller. We may, therefore, grind the horns straight, the faces being about tangent to the path of the roller pin's face; or to be more exact, the faces of the horns should be about parallel to a line tangent to the path of the roller pin's face; for, of course, the horn must safely clear the face of the roller pin when the fork stands against the bank. These faces should also be polished, in the same manner as the sides of the notch.

A VARIATION ON THE FOREGOING.

A fault frequently found in even the finest watches, is a *worn* fork; it is sometimes aggravated by incorrect form, or slight roughness of the roller pin. In many fine watches the triangular pin is used, and in the effort of the maker to economize power, the edge of the pin is made much too thin, affording insufficient surface of contact, consequently, abrasion soon takes place. But even with pins correctly formed, wear sometimes occurs. Oil is unquestionably a disturbing element, and many fine workmen do not apply it to the pin or the notch.

Wear takes place; let us consider how best to repair the injury. If the wear is but slight, it may be polished out without seriously impairing the escapement action. To accomplish this easily and in the best manner, a thin-edge tortoise-shell lap, charged with finest diamond powder is excellent. No fear need be entertained that the fork may become charged with diamond; there is positively no danger if the fork is of tempered steel; a diamond lap would not be suitable for *soft* metal forks.

To hold a fork to polish out the notch, a

K. & D. second hand holder, with the jaws filed away, as shown in Fig. 10, is very convenient; we may grasp the pallet or pallet staff with sufficient firmness; possibly needing to hold a finger against the edge of the fork, while the lap is doing its work; but this is an advantage rather than otherwise, for it prevents vibration of the fork. Care should be taken to hold the fork slightly below the center of the lap, if the lap turns toward you, and *vice versa*. As worn spots in the notch are usually much less deep than they appear, care should be taken to remove no more metal than is really necessary. Of course the guard finger must be removed.

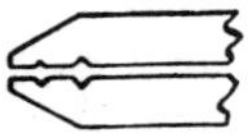

Fig. 10

After finishing the notch, a careful operator may, by laying the fork bottom up on a polished hard steel stake, peen with a small hammer along the edge of the fork just back of the notch, thus closing the horns, to bring the shake of the pin in the notch to the right amount. The greatest care should be exercised not to overdo the peening, as the effect of a light blow at the right point is very marked. All this may be accomplished in a very few minutes, with positively no material injury to the fork.

Closing of the fork by cold peening is recommended only within narrow limits, .05 mm. at most. If more is necessary, we should anneal the fork, close it while soft, then re-harden and temper. Let it be understood that we are now talking of the extremely delicate forks found in fine Swiss watches. Some may exclaim, What nonsense! to go to all this trouble to bring a notch to fit the pin, when it would be so very simple to finish out the notch and fit a pin large enough for it.

Notwithstanding, the method we propose is probably the shortest cut to best results. Suppose the watch we have in hand is a finely adjusted movement; to fit a pin even .05 mm. larger would make a difference of as many seconds a day in the isochronal adjustment; the effect would perhaps be even greater than this. Therefore, if we work for fine results, we must be careful; a seemingly short cut home is often longer in the end, than a long one. Besides, even considered from the standpoint of mechanical operations only, there may be some question whether or no, to change the pin would be the easier. In fine watches a change of pin usually necessitates grinding out the hole in the roller, the pin usually fitting closely. We will not present this as argument, however; for the reasons previously given, we should never put in a larger pin unless our knowledge of the previous rate of the watch indicates that such change would improve the isochronal adjustment.

In general, a larger pin will cause a loss in the short arcs; and it is hardly necessary to say that it is not due to any possible or probable difference in weight of the two pins; but on account of the slightly different angle at which the pin makes and breaks contact with the fork. For, as is well known by all watch adjusters: taking a balance vibrating freely, entirely detached from the escapement, it will have a certain definite time of vibration; then any abstraction of force—such as unlocking—taking place before the dead point, slows the time of vibration. And also, any addition of force—such as impulse—taking place after the dead point, slows the time of vibration. In the case of the roller pin we are discussing, other factors tend to quicken the time of vibration, but the greater number tend to *slow* it.

Now, the problem of annealing the fork; as every watchmaker knows, they are very delicate, and to him who has never tried it, the job might make him chilly. However, if we go at it right, it is really very easy; we have performed it upon hundreds of forks, with never an accident. Remove the guard finger; take a piece of thin sheet-copper or brass, about 3 mm. wide by 1.5 cm. long, and double it through the center of its length. With a thin knife blade, carefully open the ends and insert the fork up to about the guard finger screw hole. We will now hold the metal strip in the flame of a small lamp, which should be held in a slightly darkened place, to enable us to see exactly where the flame is striking; we heat the strip to a dull red, and slowly withdraw it to a short distance from the flame, and allow it to cool. We should guard against any quick movement of the heated fork, as the slight draft of air thus caused might re-harden the fork, and an attempt to close it might end in disaster.

Before inserting the fork in the folded metal strip, it is well to smear it with castile soap, or to put some soap in the crevice of the metal strip; this tends to prevent oxidation of the fork while heated. If the heating operation has been properly done, the fork is now quite soft, and may readily be sufficiently closed, without danger of breaking. A good way to do the closing is to rest the fork on a round-top stump, and with a round-edge peening punch, slightly bend each side of the fork in.

The conditions are shown in Fig. 11, in which *a* shows the stump, *b* the fork, *c* the punch; *d* shows an end view of the punch or stump. This is to make it clear that it is not simply a round-face punch that is meant. These round-edge peening punches were originally intended for stretching a balance arm, but they are more useful for other purposes. In closing a fork in this manner, the punch should be placed very near the bottom of the notch, as shown; otherwise even the slightest tap with the hammer would close it too much. Sometimes it is preferable to have the stump touch the fork at a point slightly back of the notch and the punch slightly forward of this position. By this method we close each side of the fork separately; this is the plan we generally follow, for it is difficult to close both sides at once, perfectly true. (We made a special tool for this work, with a movable jaw, controlled by a screw; it worked well.)

Fig. 11

A fork should be closed in no more than is necessary to allow of properly refinishing the notch. As the closing progresses, we should try it on the pin, thus guarding against overdoing the matter. The closing, of course, changes the direction of the horns slightly; in some modern Swiss watches, we may be obliged to correct this, but in the majority of older ones, no change in the horns of the fork will be necessary, for they are generally much too open. By this, we mean that were the guard finger pressed against the safety roller, the outer end of the horn would be much too far from the path of the roller pin. Some watches that are generally of good quality as regards workmanship are very bad in this particular. In such cases, closing the fork results in a marked improvement of the horns.

Our fork being annealed, and closed sufficiently to allow of entirely removing the worn portion, the next step is to reharden and temper it. In hardening we will make use of the same, or a similar strip of copper or brass, that we used in annealing; and we will not forget to apply the castile soap before heating. For holding the liquid in which to plunge the fork, an ordinary test tube about six inches long is excellent; it should be filled to the brim with water, which must not be too cold—about 70° Fah. is best.

Now, with the fork held firmly in some suitable implement, and with the copper strip applied, we place our lamp in a darkened place and heat the copper strip to a cherry red, and quickly plunge it in the water; the test tube should be held right up close to the flame while the fork is being heated. In fact, the chief object in using a test tube is to enable us to have the water the shortest possible distance from the heated fork, and avoid overheating or burning of the steel.

The fork being hardened, the next step is to temper it by the oil bath process, when it may be cleaned up and the notch refinished. If the foregoing operations have been skilfully performed, the fork will not be discolored beyond a point about a millimeter from the guard finger screw hole. If the fork is of a type to admit of separation from the pallets, it should be done before tempering. If fork and pallet are one piece, the stones should be removed and reset after the fork is tempered; the temperature of the tempering bath is too high for shellac; it would be burned, and the stones probably come loose in service.

Now, it is clear that at some point in this

fork, between the original hardening and tempering, and the new work which we have done, there must be a narrow portion of soft steel. To some, this might constitute a serious objection to the method. We do not think it does. In the first place, it comes in a comparatively broad part of the fork, where there is ample strength anyway; and second, under the stress which occurs when the fork is in actual operation in the watch, the fork is really more rigid at this point than it would be if hardened and tempered. To illustrate, if we were to take a rod of soft steel and another of the same dimensions, but hardened and tempered, and fasten one end of each securely, allowing the free end of each to project exactly equal distances, then, under a given force, the tempered rod will vibrate much farther than the soft one. This shows that for the purpose of transmitting impulse only, a soft fork may be more rigid than a tempered one. Of course, the elastic limit of the soft fork is much below that of the tempered one.

After tempering, the next step is to remove any slight discoloration; this may easily be done by means of fine oilstone powder, on a match splint, holding the fork on a piece of cork or fungus. Of course, the polish on top of the fork as far up as it was heated has been slightly dimmed; we will consider its restoration after we have finished the notch. In grinding the worn spots from the sides of the notch after the fork is closed, hardened and tempered, by means of the diamond lap, care must be exercised, for diamond laps cut rapidly. To be sure, we are using a very fine lap, but the work is also fine, and it is nearly always desirable to remove no more metal from the notch than necessary; therefore, we should have a very slight remnant of the "pit" to be removed by the polishing lap. The laps thus far spoken of for refinishing forks are circular laps to be used in the lathe. They are flat face and thin edge. It is not necessary to have laps of this kind, charged with diamond for this work, although very convenient. Excellent work may be done with simply a thin flat strip of soft steel, charged with fine oilstone powder mixed with oil, for grinding, and a boxwood lap, to be used in the lathe, for polishing. As the sides of the notch making contact with the roller pin should be rounded slightly, and not perfectly flat, we may make the polishing lap of such shape as to give the surface the required form. Such a lap is shown in Fig. 12, front end view and in section.

Fig. 12

This lap is made of boxwood; the one shown is with taper hole to go on the lap truing chuck of our pivot polisher, although, of course, it may be mounted in any way most convenient to the watchmaker; the essential condition being that it shall run true. The inner surface of the flange, *a*, which projects at an angle somewhat like the rim of a tin pan, is the part which we utilize for the polishing surface. The flange, *a*, must be thin enough to enter the notch in any fork; the lap should be made of a good, firm piece of boxwood, the grain running preferably, we think, in the direction of the hole. Now, after having ground the notch with the soft strip and oilstone powder, rounding it slightly in a transverse direction during the grinding process, we may, by slightly smearing the inner surface of the flange, *a*, with diamantine or sapphirine mixed with oil, and holding the edge of the notch in contact with *a*, slightly traversing it in a direction lengthwise the notch, very readily bring up a brilliant finish.

Many watchmakers may not find it convenient to provide themselves with diamond laps, but the means involved in the last-named method of finishing the notch are so very simple that no workman need be without them.

An efficient means of touching up a worn fork or fitting up a new one, is a small sapphire slip, fitted to a light handle. A diamond lap is convenient for shaping the sapphire, and giving it a surface with the right degree of cut, but it may be done very successfully with a fine carborundum hone. The sapphire should be attached to a piece of brass, by means of lathe cement, the cement being allowed to come up even with the top of the sapphire; this to a great extent prevents chipping the edge of the sapphire. The grinding should be done crosswise, and the stone kept moist with oil or water. The sapphire must be thin enough to enter edgewise any fork notch; the width may vary to suit our pleasure. The form of the sapphire should be as shown in Fig. 13, which is a cross sectional view, very much enlarged. This form enables us to finish out a dovetail notch and have the corners perfectly sharp. The sapphire should be kept slightly moistened with oil, and we may know where the cutting is being done by the slight deposit of gray steel cuttings which spread over the surface. This is very important, for the sapphire cuts quite rapidly, and unless we are careful to maintain it perfectly parallel with the surface, we are sure to get the notch out of shape. And also, if we need to correct or modify the form of a notch, it is equally important to know just where the cut is taking place.

Fig. 13

Such a sapphire need not be long, 1 cm. being sufficient. A convenient means of mounting it in the handle is to drill into the end of a piece of brass wire of suitable size, a hole about 3 mm. deep, which may then be formed to approximately fit the sapphire. A convenient way to do this is to file up a piece of steel of the same shape and size as the sapphire; harden and temper it. By driving this steel form into the hole in the brass wire, we may, by careful hammering, bring the hole to fit the sapphire, when it may be securely fastened in place by means of lathe cement; the brass wire may then be inserted in a light handle; a straight piece of pegwood answers as well as anything. Such a sapphire leaves a very smooth surface, almost a polish; it requires but a light touch with the polishing lap to flash up a mirror-like brilliancy.

We may now consider the matter of refinishing the outside of the fork after it has been annealed and rehardened and tempered. The corner of the light Swiss forks are usually finished with a sort of rounded chamfer; ordinarily this may be restored by light rubbing with a small oval burnisher; where this does not seem sufficient, a very small quantity of diamond powder applied to a flattened pegwood will quickly restore the polish. A leather buff with rounded edge made to use in the lathe is also very useful. Such a buff may be conveniently made by glueing a piece of sole leather to the face of a wooden lap; the leather may then be turned to the desired form. We may use diamantine and oil upon it, or fine diamond powder mixed with oil. We must be very careful to hold the fork in such a manner that the friction of contact will be of disengaging character to avoid danger of catching. With a lap of this kind, the finish of the chamfers is restored in much less time then required to tell about it. Such a lap is useful for a variety of other purposes, such as polishing small oval burnishers, removing specks of rust from small oval screwheads, etc.

We now come to the restoration of the top surface finish. If it is a case in which the finish is of the fine parallel line variety often seen in fine Swiss watches, the finish may be reproduced by means of a suitable diamond lap; or a quite respectable imitation may be produced by means of a fine carborundum hone. Although the carborundum hone skilfully handled produces very good work, it can only be called an imitation of the brilliant, scintillating surface produced by the diamond. The reason is very simple: carborundum, although very hard, crystallizes and fractures in a different way than the diamond. It so happens that the crystals of the diamond are of such form as to produce grooves whose sides are inclined at such an angle as to decompose the light rays, producing the brilliant prismatic effect.

Should the finish be a polish, we may readily restore it by holding it down on our block tin polishing lap, with the perfectly clean tip of a lead pencil, or a bluntly sharpened pegwood may be placed in the guard finger screw hole. Ordinarily we need only operate on the portion of the fork which has been dimmed, the remainder of the fork may project over the edge of the lap; but very small strokes will be necessary in polishing so small a surface.

There should be no evidence of patching in the finish. A small tin lap to use in the lathe, is useful for touching up the flat polish on top of the fork; it should run slowly, and the fork should be held by pressing the end opposite the notch into a piece of pitch. This allows it to yield slightly, and the surface will remain flat upon the lap where it is held by a pegwood in the guard finger screw-hole. The fork should be placed against the lap before the lap is started.

See special article on Polishing Steel.

A kind of work coming under this head, which is much less delicate than closing of forks, is the stretching of stem wind yokes and similar pieces, to correct a faulty depth or other action. Where this class of work is necessary, we have found it worth while to soften the part and do the necessary stretching by laying it on a large flat faced stump, using the peening punch; the punch should be used on the under side of the yoke or other part, and, of course, the indentations should be at right angles to the direction of the stretching required.

Operations performed in this way can hardly be said to be unworkmanlike, for when we reharden, temper and refinish the part, it is practically as strong as new, and the appearance is not marred in the least.

It may seem to some watchmakers too much work to soften a piece, reharden and temper, not to mention the refinishing. Well, we speak conservatively when we say the whole thing may be easily done in ten minutes; which *we* think is much better than to risk breakage by attempting to stretch a tempered part, not to mention spoiling its appearance.

In hardening a flat piece like a yoke, it should be plunged into the water edgewise, or endwise, to prevent springing it; it should also be carefully covered with castile soap, so it will come out a fine silvery white, without scale.

For tempering such work, by far the best method is the oil bath. For this purpose we need a metal cup, about 1¼ inches deep by 2 inches in diameter; to this should be fastened a handle about 6 inches long, and of a character that will not get too hot to hold. Different oils may be used, but our preference for watchmakers' use is olive oil. It should be heated—not too rapidly—until when held down beside the flame the vapor rising from the heated oil will ignite in brief flashes; this is hard spring temper; when the vapor ignites and continues to burn on top of the oil, it is a softer spring temper, all right for case springs, etc., but too soft for balance staffs.

As oils vary somewhat, the workman should experiment a little; the art is soon acquired, and the results are more uniform, and more quickly produced than tempering by colors. Besides the work comes from the oil-bath clean.

Now the matter of finishing a stem wind yoke deserves some consideration. A great many of them in the older American watches were finished gray. The job shop method of doing this is familiar to every watchmaker, but it isn't so absolutely simple a matter to get a good surface, without scratches, as the usual instructions seem to imply.

In the first place, the oilstone powder, as we buy it in the market, is not graded, nor is it always clean; often containing black specks, and other foreign matter, which produce scratches if allowed to remain.

A convenient way is to take a bottle with wide mouth, similar to a vaseline bottle, and about the same size, fill it about half with good sperm oil; now put in, say two tablespoonfuls of oilstone powder; cork the bottle tightly and shake it thoroughly; then set it aside to settle. The oil may then be turned off, but should not be used for lubricating purposes.

Now, when we want to do a job of gray finishing, we can get some powder from the top of the layer in the bottom of the bottle, that will be free from hard particles, or coarse grains. To gray finish, the motion should be short and circular, rather than simply back and forth, as we might do in ordinary grinding.

Some prefer a soft steel or cast iron grinding lap, but ground glass will probably be most easily procurable, and serves well; it should be kept flat and in good condition by frequent grinding.

Often the gray finish of our work will be found to be darker, or more "dead" than the other parts; the color may be lightened by rubbing with a flattened piece of pith, with oilstone powder and oil, rubbing in the same circular manner that we do when grinding on the ground glass.

Another kind of finish, more frequently seen in Swiss watches, is the straight line finish; fine parallel straight lines running from end to end. This finish is very easily produced, although it requires some care. Lay the work on a flat piece of cork, or fungus, held in the vise, and with a fine flat carborundum or India oilstone hone, take a stroke over it lengthwise, bearing down firmly the while. The clean cut parallel character of the lines depend upon how straight and even we carry the hone.

The stone may be used dry, or with oil or water, depending somewhat on the color or shade of surface we wish to produce. A little patience and perseverance will enable the workman to produce very respectable results. Of course, the *ideal* method of producing this lined surface, is with a soft steel or copper lap thoroughly charged with diamond powder of suitable grade.

The crystals of diamond being of such form as to give the most advantageous angle to the sides of the microscopic grooves; the result is partial decomposition of the light rays striking it, producing most beautiful nacreous or prismatic effects; a shimmering iridescence plays over the surface when the light or the piece is moved.

Perhaps it may be well to explain what is meant by fungus, which we have several times mentioned in the course of these articles. It is the approximately semi-circular projection seen on dead trees or stumps; often growing to the width of a foot or more; although the size varies greatly with different localities. The white portion underneath these formations is, when dried, very tough and elastic, far superior to cork to hold work for many grinding and polishing operations.

Every watchmaker should be able to produce a perfect polish quickly; nothing gives an air of crisp completeness, skill, art, and mastery of ones craft so much as a little polish perfectly done. Not a milky or foggy substitute, but the deep black polish, in which we cannot tell where the ray of light ceases and the surface begins; such surfaces always remind us of a deep, dark pool, where the brook trout congregate.

Every workman can polish if he will, and in another chapter we have described in detail, the art, as best adapted to the watch repairer's needs.

Coming back to strictly stretching operations, we have the case of forks of the Swiss style —we have already closed the notch of one— being too short to afford correct safety action, or frequently the horns are too short. Instead of smashing down the guard point, as we frequently see done, or cutting a notch through, and bending the partially separated portion forward, we soften the fork and stretch it with the peen punch in the manner described for stem wind parts. This makes a thoroughly workmanlike job that none of us need be ashamed of; to restore the finish is so quickly and easily done, after we acquire the art of polishing, that none would hesitate on that account.

What we most need in the watch repairing business is a higher ideal of *mechanical* morality; many a workman who would hesitate at wronging a fellow mortal, will most cheerfully butcher a fine piece of work.

Polishing Steel

PERHAPS no process involved in watchwork is more difficult for the young workman to acquire than the polishing of steel. It may be truly said to be comparatively an unknown art among American watchmakers; the awful substitutes for polish that some of them produce should be ample proof of this rather sweeping statement.

An English watchmaker, on the other hand, while he may be unable to handle our American tools with anything like the facility that we do, can almost invariably do a beautiful job of polishing, and that too, with the most primitive means. It is chiefly in the "know how", and cleanliness.

There is no method of steel polishing so convenient for the watchmaker, nor by which such excellent results may be obtained, as with a block-tin lap, using sapphirine or diamantine as the polishing medium. The process is very simple; but as given in books, the instructions are not sufficiently explicit at the very points where explicitness is essential; and in consequence the attempts of the beginner are usually failures, where, if rightly done, the very first attempt should bring gratifying success. We will describe the tools necessary, and the mixing of the material so carefully, that if followed, failure is impossible. We will first provide the tools.

Block-tin laps may be bought in the market, but we can easily make one much better adapted to our purpose and costing less. Tin is not easy to work with a file, and when our lap needs truing we do not want to send it to a machine shop to be planed. Let us take a brass disk 3 inches in diameter by $\frac{1}{8}$ inch thick (see Fig. 1); in the center fit a threaded brass plug fitting the screw chuck of our lathe. This plug must be solidly fitted; best done by tapping the disk with our cement brass tap and screwing the plug in tight; when screwed in the chuck, the disk should come snugly against the chuck. We will now fit three sturdy legs, projecting from the back of the disk slightly more than the threaded plug. These legs

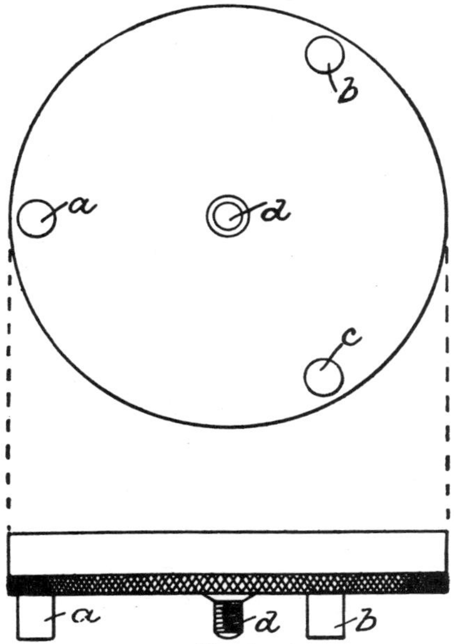

Fig. 1

should be as near the edge as practicable and equally spaced around the disk. They should be ¼ inch diameter and may be screwed or riveted in. The disk, when the plug is screwed in the chuck, should run true or nearly so.

The side of the disk bearing the legs is the bottom. We will now brighten the top with sandpaper, or in any convenient way, and apply a thin coating of soft solder; and a convenient way to do this is to apply the soldering fluid; apply the solder, and heat the disk over a bunsen flame until the solder melts, when, with an old brush, used wet, the solder may be spread over the surface; the exact thickness of the solder coating is not important.

Now apply a coating of hard soap to the edge of the disk and bind a strip of strong paper around, reaching about ¾ of an inch above the top. The paper should go around several times and be securely bound with string or wire. We will now melt about 1½ pounds of block-tin and pour it within the paper wall; the disk resting upon its legs and a level surface. The object in applying the soap is to stop any small crevices, thus preventing the molten tin running through. Before pouring the tin, skim off any dross or other substance floating upon it. It is well to have the disk warm, but it need not be hot. As the solder melts at a lower temperature than pure tin, the tin when cool will be found solidly attached to the disk.

We will now remove the paper wall, and screwing the plug in the chuck, we can face off our lap in comfort; we will also smooth the edges and if desired the edge of the brass disk may be knurled, and we will have a lap better than any we can buy, and one we can keep true without trouble. Any finishing desired on the bottom of the disk should be done before facing the lap. This is to avoid imbedding any grit in it. This lap should be kept covered perfectly in a box free from dust; don't try to keep it by wrapping in paper or cloth. When not in use, the tin surface should be free from contact with everything but air.

To keep the surface of this lap in the very best condition—between the light turnings sometimes required—we need a scraper. In the use of a scraper, such as we are about to describe, we are departing from the beaten path of those skilled in this art, and making a decided improvement at the same time. Scraping is least likely to imbed any particle of grit, and the metal removed is a mere film; therefore, our lap remains true for a long time, and the surface left by the scraper is the best possible for the reception of the polishing material.

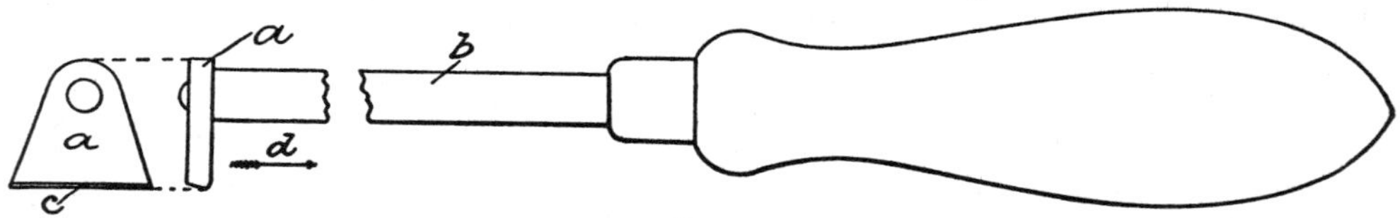

Fig. 2

The scraper is shown in Fig. 2; the head, *a*, is made of steel about $\frac{1}{8}$ inch thick; the edge, *c*, should be about $\frac{3}{4}$ of an inch long, hardened and tempered light straw, and ground very straight. The angle forming the cutting edge should be about as shown. The rod, *b*, joining the scraper head to the handle may be of brass or steel $\frac{1}{4}$ inch diameter; an ordinary file handle may be used, and the total length, including scraper, should be 7 or 8 inches. In using the scraper, it should be drawn in the direction of the arrow *d*, Fig. 2. In using our lap it is best not to use the same spot all the time; the frequent scraping will soon produce a low spot and the lap will require facing in the lathe much sooner than it will if we use judgment in the use of the scraper.

In addition to the tin lap, we need two glass laps 3 or 4 inches in diameter and about $\frac{1}{4}$ inch thick. Sometimes the heavy glass used in some French clocks can be utilized. The glass laps should be carefully roughened with emery or carborundum; No. 120 emery is suitable. The surfaces should be true and flat. One of these laps we will use with oilstone powder mixed with oil and the other with diamantine or sapphirine.

To hold the work we use the well-known tripod or screw-head finisher, illustrated in any tool catalog, also shown in Fig. 3. The ends

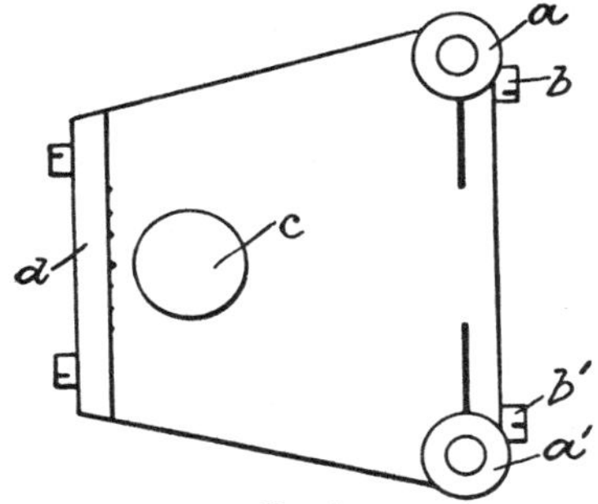

Fig. 3

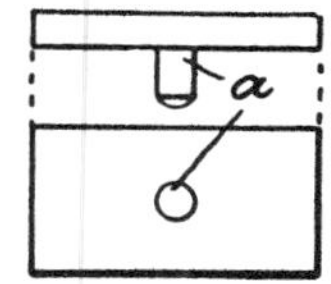

Fig. 4

of the two leveling screws should be hard and the screws should move with sufficient friction in their holes to prevent accidental shifting. The corners of the screw ends should be rounded to prevent cutting the surface of our tin lap. Under the clamp of these tripods there are usually several notches of different sizes, to facilitate the holding of screws, etc. The middle one is usually the deepest, if not we will make it so and of a size to conveniently grasp a piece of wire $\frac{1}{8}$ inch diameter. We will now prepare several brass plates (Fig. 4), $\frac{3}{32}$ inches thick and of sizes convenient to hold such articles as we will need to polish, such as regulator springs, endstone caps, etc. From the center of these brass pieces projects upward a steel or brass pin, *a*, $\frac{1}{8}$ inch diameter, to fit the large notch in our tripod. We also need a three-storied polishing block; shown in Fig. 5.

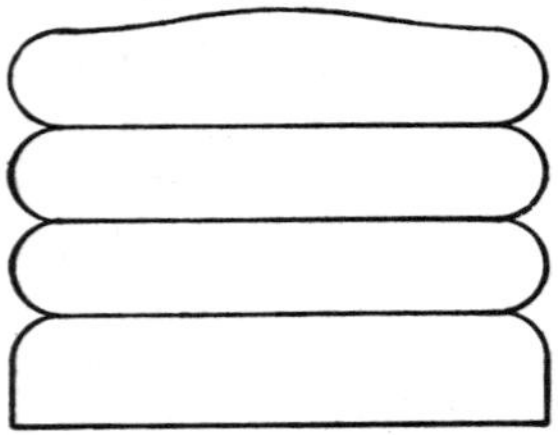

Fig. 5

Having the necessary tools, now we will prepare our polishing material and do a job of polishing. We will try sapphirine, and as the quality is not uniform, and some samples may contain grit, etc., which would spoil our job, we will proceed on safe lines. Take a wide-mouth bottle about the size of a five-cent vaseline bottle—it should be clean—fill it about two-thirds with benzine; empty the contents of a bottle of sapphirine into the benzine, stirring it up thoroughly; plug the mouth of bottle loosely with jewelers' cotton and set it aside to settle; leave it until the benzine is all evaporated—the cotton effectually excludes dust but allows the evaporation of the benzine. Any grit or coarse particles will now be at the bottom of the bottle. Until we have mastered the art of polishing we will be careful not to go more than half way down through the deposit of sapphirine in the bottom of the bottle. We may eventually find that all of the sapphirine may be fit to use, but we will take no chances at first.

Now, on the top plate of our three-story polishing block—which we will wipe off carefully with our *hand*, to be sure there is no grit on it—we place two drops of watch oil. With a clean knife blade, we take from our sapphirine a lump about the size of a pea and add it to the oil, mixing it carefully and adding more sapphirine or oil as necessary until we have it mixed as stiff as possible, say to the consistency of *cheese*. And right here is where most beginners "fall down" in trying to follow the instructions usually given: they do not mix the material *stiff* enough. So do not make a mistake; you cannot get it too thick, so long as there is no *dry* sapphirine present. We will take the bottom plate of our polishing block to mix the oilstone powder on. As we are doing a "dead sure" job this time, we will treat the oilstone powder just as we did the sapphirine. The oilstone powder must be mixed quite thick also; about like good putty.

Just one more item, and we start on our job: this is what an English finisher would call "watchmakers' putty." To prepare it, take a piece of bread, 2 inches square, or thereabout, without crust; if at all dry, moisten it with water and squeeze the water out as well as possible, working the bread up into a lump; stick it on the end of a wire and toast it over the bunsen flame until a light brown on the outside; knead it over again in the fingers and repeat the toasting as before until the "putty" thus formed, while soft, does not stick to the fingers. It usually takes about three toastings, but much depends on the bread in the first place, and how well you squeeze the water out.

Now for our job. Let us polish a regulator spring. First fasten it to one of the plates, Fig. 4, using lathe wax or shellac, and no more than is necessary to attach it firmly at every point; no part should be left unattached, as it may spring more or less and interfere with the success of our job. We will now fasten the plate in the tripod firmly, letting the back of plate rest against the bottom of the tripod. Before beginning to grind with the oil-stone we will level it as nearly as we can by applying the tripod with the work against the surface of our glass lap, sighting through under it and turning the adjusting screws as required, until as near as we can judge the surface coincides with the glass plate. Now apply a thin coating of our oilstone mixture, as thin as we can spread it with a knife blade; do not cover the whole lap; a place as large as a half-dollar is enough.

Now, while pressing the work gently down, proceed to grind it by small circular strokes; do not proceed far before you examine the work, to see where the grinding is being done, and adjust the screws if necessary to perfect the parallelism. When the surface is of a uniformly gray appearance all over we will remove the oilstone powder by washing with benzine, followed by soap and water. Don't be afraid to get the tripod wet, the whole thing should be well washed and rinsed; getting it perfectly dry is not important, the water can be shaken off, or carefully wiped; care being taken not to move the adjusting screws.

We will now apply a very thin coating of sapphirine to the *glass* lap, which we intend to use for that purpose. You will understand that after grinding these glass laps with the emery, they are to be carefully washed. A brief grinding on the glass lap with the sapphirine brings up a surface ready for the final polish. Clean off with benzine, and dab it repeatedly with the lump of watchmakers' putty; this lifts out any loose particles of grit which may remain after washing. Examine the work carefully with a glass, and if the shellac is anywhere near level with the ground surface, remove it before attempting the final polish; this may be done with a small triangular brass scraper, Fig. 6.

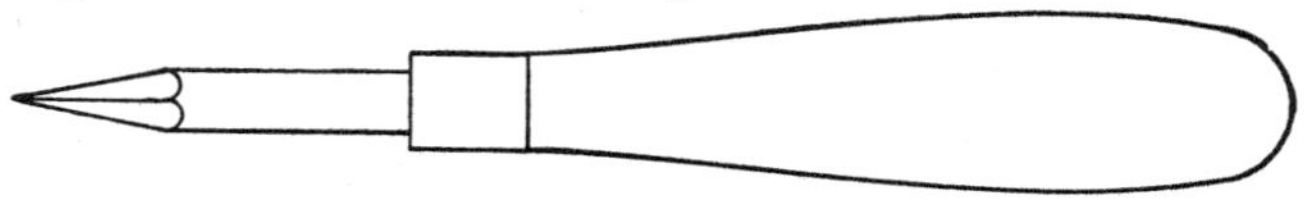

Fig. 6

Now for the tin lap. Apply a dim of sapphirine to the surface about as large as a quarter; "dim" describes the amount very well. It is best applied to the surface with the first joint of the thumb, (Fig. 7), by touching the sapphirine "cheese" lightly and dabbing the tin lap. If you breathe upon the end of a clean finger and touch it immediately upon a clean mirror, the resulting impression will give you an idea of the "dim" of sapphirine to be applied to the tin lap.

Fig. 7

After applying it, carefully use the bread putty on the work again, and on the ends of the adjusting screws, and apply the work to the "dim" on the tin lap; rub it back and forth as nearly in straight lines as convenient, bearing down lightly, and being careful not to let the tripod tilt and dig a rut in the lap. A few seconds rubbing will flash up a polish such as you see in fine watches; and what workman has not admired it? This process may sound tedious, but it takes actually but a short time to do the work.

The art once learned, you will have no further use for Vienna lime or any other sloppy method of polishing. After the "knack" is once acquired it will not be found necessary to use the bread on every little job, such as end-stone caps, screw heads, etc. Frequently on end-stone caps a good polish may be quickly had by simply laying it on the lap and moving it by means of the clean rubber tip of a lead pencil. Dip the rubber tip in benzine and wipe it dry with a clean cloth.

Many small articles, such as regulators, may be quickly polished with sapphirine by laying on a cork held in the vise, first grinding with a strip of ground glass and oilstone, or a flat Arkansas stone slip, cleaning carefully, also a clean cork and applying a bell-metal strip with a "dim" of sapphirine. The bell-metal strip should be filed flat and stoned smooth with a Scotch stone; clean carefully and finish with a tin strip. A tin strip may be made for this purpose by melting tin onto a brass strip, as described for the lap. It may be provided with a handle like a file, Fig. 8.

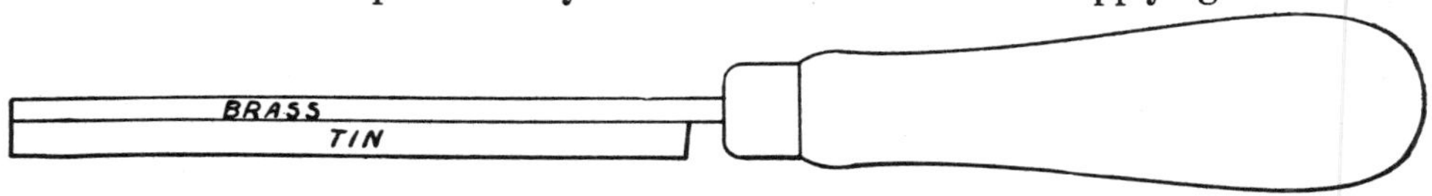

Fig. 8

Diamantine may be used the same as sapphirine, with satisfactory results, but it is not quite as quick in its action. It will be noticed that it is recommended to move the work on the tin lap as nearly as possible in straight lines. This is correct. You will, I think, find that a deeper black polish is produced than will be the case if the work is moved with a circular stroke. It is true that a polish when highly magnified is seen to consist of exceedingly fine lines crossing and recrossing, but these lines are of such fineness that they are crossed even when we move the work as straight as possible by hand. Anyway, if you follow these instructions success is yours.

A small spatula, shown in Fig 9, is very convenient for mixing the polishing material. It consists of a piece of brass wire $\frac{1}{8}$ inch in diameter by about 3 inches long, with a slit sawed in one end, and a wide piece of mainspring, *a*, inserted and soldered in place. The end of the steel blade should be somewhat curved, as shown, and ground down quite thin, but not sharp. Do not use the same spatula to mix the oilstone powder. Every precaution must be taken to avoid the presence of grit of any kind in our polishing material.

Fig. 9

A very convenient case for our tin lap is shown in Fig. 10, the cover being shown in section. It is simply a wooden block turned as shown, the upper surface being provided with holes to receive the screw plug and legs of the lap. The groove, *d*, in the base is to afford a hold for the fingers in lifting it. The groove, *c*, is for the reception of a few turns of common cotton string which completely excludes the dust when the cover is applied. The top, *a*, of the cover may be made of wood and the sides, *b*, of several thicknesses of tough paper glued together, or the cover may be made of metal. When we have such a case, in using the lap, we remove the cover and leave the lap in position on the wooden block, as it affords a stable base, and is convenient

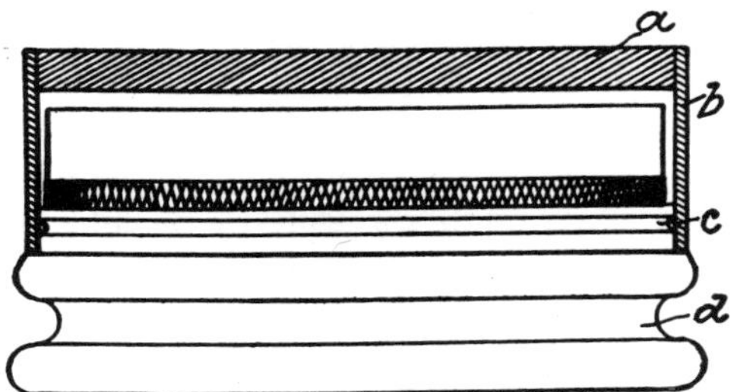

Fig. 10

to handle. A substitute for the scraper shown in Fig. 2 may be made by softening the end of an old 5 inch flat file and bending about ½ inch of it at right angles; file up the angles as shown, harden and temper.

Fig. 11

In finely finished steel work, involving screw holes, there is usually a round-bottom recess around the hole, as shown in Fig. 11. To be able to finish these quickly, yet perfectly, is very desirable. It should be the ambition of every real watchmaker to be able to finish his work equal to the very best. To polish these round-bottom recesses with the utmost ease, we will need the simple arrangement shown in Fig. 12.

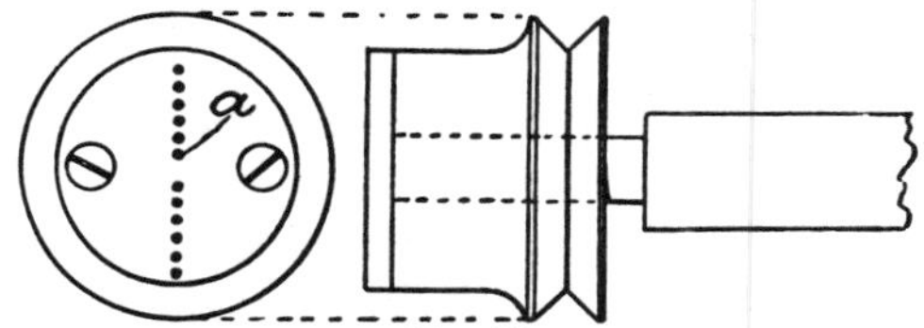

Fig. 12

It consists of a steel center fitting the tailstock spindle of our lathe, the projecting end being about ⅝ of an inch long and turned down straight to about ⅛ of an inch in diameter; the end should be hardened and ground smooth.

A pulley made of brass, and a running fit, goes on the outer end of the steel center. The hole goes entirely through the pulley, the outer end of which is covered by a steel disk, having round-bottom countersinks in its outer face. These countersinks are disposed at varying distances from the center, as shown at *a*, Fig. 12. The steel disk may be attached to the pulley by soldering or by screws, as shown. In making a new piece of work we would countersink the holes when the steel was in

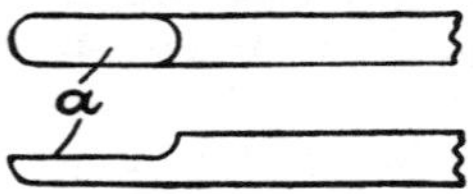

Fig. 13

the soft state. The wheel countersinks which may be bought, are suitable for the larger sizes; for very small holes we may make them of the shape shown in Fig. 13, which is simply a piece of steel with the end rounded, hardened and tempered and then slightly less than one-half the diameter ground away as shown at *a*. Countersinks made in this manner do perfect work.

The laps used to grind and polish out these round-bottom recesses are pieces of brass wire about $\frac{3}{4}$ of an inch long, rounded at one end,

Fig. 14

and pointed at the other, as shown at Fig. 14. Of course, the rounded end, *a*, must fit the recess we intend to polish. The pulley is driven by a small belt or cord from the countershaft. We will now proceed with a job. The work is cemented up in a lathe, with the recess we propose to polish centered. We then apply a small quantity of oilstone dust and oil to the rounded end of a suitably-sized brass piece, Fig. 14. With the tailstock in position, carrying the pulley, we move the spindle forward until the pointed end of the brass piece rests in one of the eccentric round-bottom countersinks in the face of the steel disk on the pulley.

In the beginning we will use a hole quite near the center. The belt from the countershaft, which drives the brass pulley, should be crossed so the motion will be opposite to that of the lathe spindle. When we have the brass piece in position, we gently press the tailstock spindle forward, while we impart motion to the work and the pulley. It will be seen that as the outer or pointed end of the brass piece is carried in a circle around the lathe center, and the rounded end remains at the center, the result is a movement of the rounded end in the recess we are to finish, which rapidly grinds it out smoothly and of perfect form.

When ground to our satisfaction we will wash away the oilstone powder with benzine;

and with a knife we scrape the rounded end of the brass piece, to remove any oilstone dust that may be imbedded therein; apply a little sapphirine and oil, and proceed as before. A very few seconds is sufficient to bring up a satisfactory polish. It is not necessary to use such exceeding care in this operation as in flat polishing. Unless the recess is very large, a third or fourth rate polish on a concave surface really looks first-class. A small piece of tinfoil pressed down with the fingers over the rounded end, *a*, of the brass piece, Fig. 14, serves a good purpose when we come to the sapphirine, on jobs where we desire a really excellent finish.

Those having a pivot polisher may fix the steel disk to the end of a lap. All things considered the little device described is most convenient, as the tailstock spindle is always on the center line; and where we use the pivot polisher we have to adjust the spindle on the center line. If the tailstock spindle moves very hard, it is best to wipe it dry, also wipe out the hole with a strip of chamois skin, by drawing it through; closely fitted sliding spindles move easiest when dry. Keep them oiled when not in use.

Staking the Balance Staff

ABOUT the simplest staking operation to be found in watch work is staking the balance staff; that is to say, it is fundamentally simple; the conditions being correct, the riveting operation is of the very simplest character. Imperfect conditions is the stumbling block to many; the staff does not properly fit the hole; there is not enough stock projecting through the arm, to make a secure rivet; or, there is far too much; the part to be riveted is not sufficiently undercut, etc., etc.

Let us first consider the replacing of a balance staff, where we do not make the staff. The cases are rare where the part of the staff that goes in the balance should not be changed in some way; generally the riveting portion is too long. But let us first get the broken staff out without spoiling the balance arm. Do not, under any circumstances, drive the broken staff out without turning away the riveting so completely that the staff can be pushed out without distorting the balance arm or throwing the rim out of true in the slightest degree.

This is not for the reason that it would be difficult to true the balance, or anything of that sort, but because forcing the riveting through the hole tears away more or less of the comparatively soft metal of the arm, thus destroying the truth of the hole, and it also enlarges it by stretching the metal of the arm, making it oval or elliptic in form.

This makes it impossible for a new staff to fit properly, and is the beginning, generally, of a whole train of evils. Let us avoid all the trouble by turning away the riveting in the first place, as we should. There are several ways to do it, and whichever one we adopt, we must take due care *not* to turn away any of the balance arm.

Every watchmaker has discovered that American staffs are very hard to turn or cut with a graver. This is because they are made by a quite different process than the hand-made staffs. Factory-made staffs are turned out complete from the soft steel, then hardened and tempered; the requisite finish being given to the pivots, etc., afterward.

On the regular factory staff, absolutely no turning is done after the staff is hardened and tempered. This method of manufacture makes it possible, even desirable, to leave the temper of the staff harder than it is practical to make hand-made staffs. When we say "practical" we mean just that. A staff *can* be turned as hard as a factory staff, but it isn't practical—takes too much time.

A few lines back we said it was desirable to have the staffs harder than hand-made ones. The reasons are: the pivots wear better; they are less easily bent. On the other hand, if *bent*, they are not so easy to *straighten;* but this, we think, can hardly be considered a disadvantage. A pivot once bent can never be as good as before, although it may *serve*, and in many cases, well enough.

For our own part we would prefer to have a pivot break square off before bending too much; it would be less likely to crack the jewel, and if bent very much, we wouldn't want to save it anyway, for the reason that it can never be restored to *perfect* form, without reducing it. The differences are not large enough to be shown on a foot rule, but they exist, and with proper instruments, can be clearly shown.

To our mind the balance staffs which are driven in without riveting are a great step in advance; the idea is by no means new, but the originator arrived too soon; it took the world a great many years to catch up with him. With these staffs, a thousand could be inserted and removed from a balance with no injury to the balance, and practically no wear, for the walls of the hole are of tempered steel, and much greater surface than the regular kind.

True, a thousand staffs may be put in a regular balance, without injury, but it requires a higher grade of skill, and *more time* to do it; there's the point. Not that a high grade of skill is undesirable—far from it; but every unit of nerve force saved on the details of a job leaves the workman in just that much better form to apply the finishing touches to the job; the touches which count so much in the performance of the watch.

To get down to the job of turning away the riveting; we will first do it by cutting out or enlarging the undercut, *a*, Fig. 1. This, on American staffs, is generally very shallow. A better way than attempting to go down with a narrow pointed graver without touching the collet hub, *b*, is to turn away as much of the collet hub as we need to, to allow us to get down with a stronger graver, and make a comparatively wide groove as shown by dotted lines in Fig. 1. This allows us to remove the whole of the rivet, so the balance may easily be removed from the staff with the fingers.

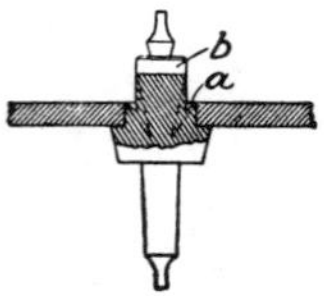

Fig. 1

No smooth turning should be attempted; in fact, a graver sharpened on a rather coarse stone cuts these hard staffs more readily than one with a very smooth edge. It takes a good graver to cut these American staffs, but good gravers can be had, or easily made; anyway, never give up with less than actually cutting the staff away as we have indicated.

Another method, preferred by some, is to grasp the staff by the collet hub in the chuck, and turn away the whole hub of the staff upon which the balance arm rests, removing the balance over the lower pivot. This is shown in Fig. 2, where dotted lines show the part of the staff to be turned away. Properly done, either of the methods outlined give perfect results. In either case, if the old staff is to be utilized for measurements in choosing a new staff, the measurements should be made before the staff is removed; except of the part fitting the hole in the arm.

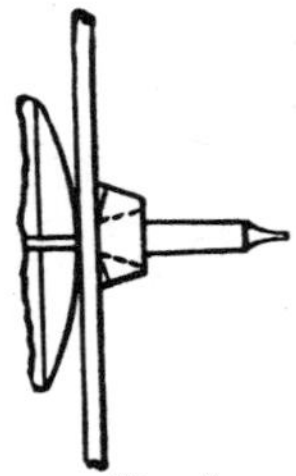
Fig. 2

Well, assuming that we have removed the old staff, we must select a new one. The only dimensions we will concern ourselves with here is the part entering the balance arm;

sometimes called the "waist", possibly because the arm encircles it. This part should fit the hole tightly enough to support the weight of the balance, when held up by the staff. It is a mistake to drive a staff very tightly into the arm; although it may go in somewhat more tightly than shown by the test we have mentioned; this is given to show what might be taken as a minimum. In any event the staff should not fit so tightly as to prevent it being inserted with the fingers, up to fully three-fourths the way to the shoulder. We should determine by measuring whether the bearing is of sufficient length to reach through the arm, and allow a sound riveting. This allowance for riveting will vary with different cases; depending somewhat on how much the hole in the balance arm is chamfered; but in no case should it exceed .1 mm., or about .004 inch; generally .002 inch is sufficient.

With a workman skilful with the graver, if the riveting allowance is too great, it may be cut down to the right length after the balance is pressed home on its seat. And in case we have to do this, it is well to cut the groove deeper than is usually done in factory staffs. Let the conditions resemble those shown in Fig. 3.

Fig. 3

Now, to the riveting; we select a hollow round faced punch, which will just *freely* go over the collet hub; the nearer it fits the better, but it must be *free*. We now select a hole in the die, in which the lower part of the staff fits as closely as possible, and yet be free; center it carefully, and bind the die securely. Now with the balance and staff in position, and the round faced punch brought down upon the rivet, as shown in Fig. 3, we hold down firmly on the punch, which is held with the forefinger and thumb of the left hand, and tap the punch lightly with the brass hammer, turning the punch slightly —say one fourth turn—after each blow; continue until the rivet is well turned, which may take six or eight light blows. Then with a flat faced punch fitting freely like the round faced punch, and held in the same manner, a few light blows completes the riveting; the rivet will be flush with the balance arm, or very nearly so, and will be almost polished; the balance will be perfectly firm on the staff, and very nearly true, often quite so.

In the riveting operation, some prefer to turn the balance and staff slightly after each blow, rather than the punch, as we have directed; still others turn both; this is a matter of personal preference, and equally good results may be had either way. It *is* better to change the relative angular position of the punch and staff, after each blow, however, no mistake about this, but just *how* it shall be done, each workman may decide for himself.

There is a class of cases, unfortunately too common, where the hole in the arm has been stretched, or at any rate, is slightly too large for the regular staffs; by "too large" we mean that the staffs would not be tight enough to meet the requirements we have given as correct. Now, if the workman has to use such a staff, it is well to know how to rivet it with the least chance of the balance being found eccentric after the riveting is done. Mr. Bernard C. Husband, a skilful watchmaker, of Boston, informs the writer that most satisfactory results are secured in such cases by using the flat-faced punch first, in fact to do the entire riveting with the flat-faced punch.

In fine staffs, after riveting, we always like to touch up the undercut and riveting slightly, with very fine diamond powder, used on a piece of ivory, boxwood, or even pegwood; it gives it that crisp "better than ordinary" appearance, which we all like to give our work.

The Waltham Taper Shoulder Detachable Balance Staffs.

These staffs are illustrated on page 68 of the catalog section. In driving out the old staff, care should be taken to see that the hub in the balance arm has a secure seating in the taper mouth of the stump—the balance arm must *not* touch the stump; the hub should never be loosened in the arm.

Notwithstanding the accuracy with which the staffs are made, it will be found sometimes that the part marked "staff" in Fig. 1, page 68, is too tight a fit for the hole in the hub. In such cases, the staff should be reduced with oilstone powder and a strip of soft steel, used by hand, while the staff is revolved in the lathe; or a pivot polisher lap may be used. The hand method is generally satisfactory, for the reduction required is but slight.

In driving in the staffs, care should be taken to seat the taper shoulder firmly; this requires a special punch, with outer diameter slightly less than the taper shoulder of the staff; see Fig. 3, page 68. These cuts are self-explanatory, and if the young watchmaker will bear in mind the points we have mentioned, he should have no difficulty.

Staking the Roller.

In all the higher priced and more complete K. & D. staking tool sets, are stumps, designed especially to support rollers while driving the staff into them. In our own practice we have preferred a small special stump for all kinds of rollers, both single and double. This stump is shown in section in Fig. 4. As will be noticed, it supports the roller closely around the hole; the edge coming in contact with the roller is rounded, making it suitable for double rollers, with polished concave in the lower face; in such cases, we prefer to lay a piece of tissue paper over the stump, although there is no danger of scratching the work if the stump is kept properly polished.

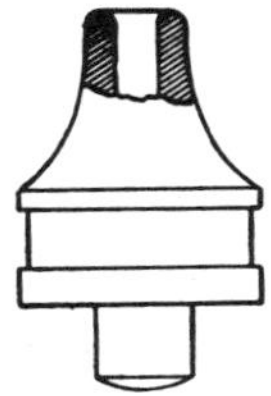

Fig. 4

With INVERTO, a suitable round faced hollow punch may be used for this purpose. We would advise every workman to have such a stump—if he hasn't INVERTO—for it is a most useful thing, in addition to staking rollers, to rest a roller on while pressing the hairspring collet to place, perfectly level. Many workmen use a pair of tweezers to press the collet home; sometimes the tweezers slip, and the hairspring is damaged—it don't pay to take such risks.

The use of the stumps, supplied with Staking Tools for supporting rollers, while driving the staff, is so very obvious, as to render any detailed explanation unnecessary.

It might be remarked however, that in cases where the staff is slender above the collet, it is wise to use a hollow flat face punch to drive the staff, rather than a conical mouth punch, resting on the cone of the top pivot. The hollow flat face punch should be of such size as would be used to rivet the staff, in short it should rest on the riveting instead of on the arm of the balance.

Some watchmakers prefer in staking rollers, to drive the roller onto the staff instead of driving the staff into the roller. Punches are supplied for roller driving, and the workman may suit his own preference. We think, however, that unless one has *Inverto*, it is better to drive the staff into the roller, for the reason that there is generally not a sufficient range of sizes in hollow stumps to properly support the staff by the riveting while the roller is being driven.

Instead of using a stump to rest the roller some workmen select a hole in the die that will just receive the staff and roller pin, then center the top of the staff with a suitable punch and proceed with the driving.

To our mind, the special stump shown in Fig. 4 beats the whole outfit; it covers the entire range of rollers of all styles, and does the work perfectly.

Removing Rollers.

The use of the roller remover stumps also is too obvious to require any special instruction; excepting, possibly, the special stump No. 305, page 71, catalog section. This stump is very useful for roller removing, and can be used for practically all styles and sizes. The roller should be pressed up the taper opening—with lower pivot up—until there is but little shake; then the staff is centered under the punch; this is accomplished by moving both the die and the stump until the combined motions bring the staff central. It will not do to center the hole in die, in which the stump fits, except in a few cases, where the size of roller is such as to bring it at a point in the slot directly over the center of the plug.

Grading Diamond Powder

LET us now consider the grading of diamond powder. We will assume that the watchmaker has the powder, and the problem is simply to grade it uniformly. Diamond powder may be bought in the market; but the cheaper and perhaps more satisfactory way is to buy the bort and crush it ourselves. If the watchmaker desires to do this, he will be able to buy diamond mortars for crushing the bort; we shall, therefore, not consider the making of a mortar.

We shall not pretend that the method of grading we are about to describe is the *best* method; we are quite sure, however, that for purposes of the watchmaker it gives good results. One carat of diamond powder is a convenient quantity to work at one time; it is small in bulk, but if properly used will do an immense amount of work, and last a long time.

We first need a bottle or two of good clock oil; the quantity necessary varying somewhat with the size of the dishes employed. In our experience in grading, we have used only Kelley's oil; we deem it best to state this because a different oil would in all probability bring a slightly different result. We also need six dishes, like a small saucer, of porcelain, or some smooth, hard material. These dishes should be about three inches in diameter; or at any rate, one of them should be about this size, the others may be smaller. Small dishes are made which are very suitable for this purpose; they are shaped somewhat like a saucer, but have a small open spout at one side to facilitate the pouring of any liquid; they come in "nests", *i. e.*, each one is enough smaller than the preceding to set within it.

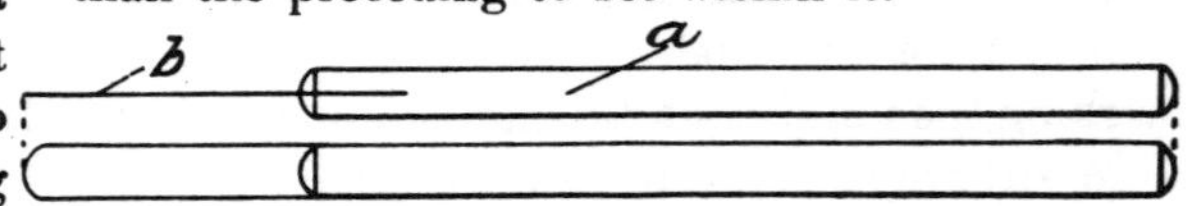

Fig. 1

We should also have a steel spatula, shaped as shown in Fig. 1, about actual size. It consists simply of a piece of brass wire, *a*, about ⅛ inch in diameter by 2¾ inches long, split with a saw at one end, into which the tempered steel blade, *b*, is soldered. This soldering should be very carefully done to leave no crevices in which particles of diamond may lurk. The steel blade is conveniently made of a piece of watch mainspring of suitable width; the point should be slightly rounded, as shown, and ground down to a thin edge.

It is advisable to have two of these spatulas, for after our diamond powder is graded, we should not use the same spatula to handle all the grades when transferring it from the grading dishes to the bottles, or when taking it from the bottles to use for any purpose, there would be danger of getting particles of the coarse powder mixed with the fine. If one wishes he may have a separate spatula for each grade, marking the number on the handles by grooves, or in any way to suit his fancy or convenience.

It is a good plan to keep the blade of these spatulas, when not in use, thrust full length into a piece of pith, using a separate piece for each one, and taking care to return each spatula to the same piece of pith every time. In short, every precaution must be observed to avoid any mixing of the powders; a very few particles of any of the coarse powders introduced into one of the finer grades will undo all our work of grading.

Now to the work of grading. We empty a bottle of oil into the larger dish; it should be nearly ⅜ inch deep; and to get this depth we may be obliged to use more than one bottle. We then empty the carat of diamond into the oil, and stir it up thoroughly with the spatula intended for use with the coarsest powder. We then allow it to settle for ten minutes, and pour off all except the coarsest particles—which have sunk to the bottom of the dish—into the next dish. This powder left in the first dish we do not designate by number; it is useful for any comparatively coarse grinding operations, some of which we will mention later.

The contents of the second dish we allow to stand for two hours, when all remaining in suspension is poured into the third dish. The powder that has settled in the second dish, and remains after pouring, we call No. 1. Unless we need the powder for immediate use, we should cover the dish and allow it to stand until the remaining oil is entirely clear, when it should be poured off, and the diamond transferred to a bottle by means of the spatula. To clean up the dish nicely is a task requiring some patience, but patience is supposed to be the watchmaker's "long suit"; besides, even the smallest particles which we can scrape up do a surprising amount of work, as the workman will soon learn. The plan of allowing the oil remaining after pouring to settle clear before attempting to transfer the diamond to the bottle should be followed through all the grades if possible. It is best to have but little oil mixed with the diamond in the bottles.

The contents of the third dish we should allow to settle for four hours, when we pour all unsettled into a fourth dish, and allow it to stand for twelve hours. We again pour into a fifth dish, and allow it to stand for twenty-four hours; then pour into a sixth dish, which should stand until the oil is clear; which may be from a few days to as many weeks, much depending on the purity of the diamond powder, for, sad to relate, the commercial article is sometimes adulterated.

The powder remaining in the third dish after pouring we call No. 2, that in the fourth dish, No. 3, fifth dish, No. 4, sixth dish, No. 5. The No. 5 is exceedingly fine, and is used on special jobs requiring a particularly fine finish. The No. 4 is the grade we will use for almost all work, such as finishing pallet stones, polishing holes, etc. Properly used, it produces a very fine surface. The bottles used to hold the graded powder is not a matter of indifference. Small ones with straight sides, see Fig. 2, *i. e.*, no neck like an ordinary bottle, are perhaps the best. They should be $\frac{3}{4}$ inch high by $\frac{1}{2}$ inch in diameter. The number corresponding to the grade of diamond may be marked on the cork, also on a small label on the bottle. These little bottles

Fig. 2

have fairly smooth bottoms inside with the corners slightly rounded; it is easy to remove the last visible particles of diamond by means of a spatula such as we have described.

Staking Train Pinions

STAKING train pinions, where a smooth arbor fits friction tight in the wheel, is a simple operation, although sometimes giving the workman considerable trouble. Frequently the hole in the wheel has to be opened slightly; this often gives trouble. If the hole has to be opened, the wheel should be chucked in the lathe, and the opening done with a small cutter in the slide rest, or jeweling rest.

If the workman insists on taking a short cut, and opening the hole with a broach, it's well to know how to do it with least chance of putting the hole out of true: in broaching, turn the broach in one direction only; not back and forth, as generally done with cutting broaches. Just how much smaller the hole should be than the pinion arbor, depends upon several things, which vary in different cases. If the wheel is hubbed, thus having a longer hole, it will not need to fit so tightly, as where there is but the thickness of the wheel to depend upon. To drive a new pinion so the wheel will occupy *exactly* the same position as to height that it did on the old one, is made extremely simple and certain if we use the Micrometer Stop. See the article on some of the various uses of that useful device.

Staking pinions of the Swiss type is a class of work where we see some awful butchery displayed. A fruitful source of this, is because it is attempted to force the pinion into the new wheel without first carefully turning away the old riveting. This should be done after the pinion is removed from the old wheel. The shoulder and leaves should be restored to as nearly as possible the condition of a new pinion. Generally it will be desirable to smooth out the undercut, bringing the end of the leaves once more to a point, or nearly so.

Then in fitting the new wheel the hole should be opened very carefully until it can be pressed on the pinion, about the same as a balance should fit its staff. If the hole is made too large, it is very difficult to rivet the wheel securely. On the other hand if the hole is too small, it is almost impossible to put the wheel on centrally; the sharp ends of the leaves will almost invariably cut more from one side of the hole than the other.

In the early days train wheels were "rounded-up" after being staked on the pinions. Under these conditions they were frequently, in fact generally, forced into holes considerably smaller than the wheel seat of the pinion. Being rounded up after staking, to stake them central was considered of small importance.

Another source of trouble in staking an old pinion in a new wheel, is that the new wheel is frequently thicker than the old one, and the riveting ends of the leaves do not reach through. To do a secure job of riveting, the points of the leaves *must* come through, and if the shoulder cannot be turned back, or the wheel made of the required thinness all over, then the wheel must be reduced in thickness immediately around the pinion; this may be on top or bottom, as the position of the wheel, relative to the other wheels, or other parts of the watch may determine.

Such reduction of the wheel in thickness, immediately around the pinion, may be very neatly done by means of a large size disk counter-sink, of the kind shown in Fig. 1.

Fig. 1

For the workman who does but a limited amount of pinion work, we advise filling in between the leaves with lathe cement; it will save him many a bent or broken leaf, besides it makes them much easier to turn smoothly. For work on pinions you *must* have a sharp graver, and it must be *kept* sharp.

In many cases some of the leaves are broken away at the part where they should be riveted. Sometimes we can get over this difficulty very neatly by turning the old wheel seat entirely away, hubbing the wheel, and staking it on American style.

Increasing thinness of watches, among other reasons, seems to be bringing the American manufacturer around to the Swiss style of inserting pinions. The same precautions must be observed in fitting as with Swiss pinions.

Fitting a pinion, Swiss style, is a job which frequently comes to the watchmaker, and a few words on the points which usually give trouble may be welcome to some of our readers.

We will assume that we are to use pinions of fine quality and temper known as Jurgensen pinions. They come with the leaves finely finished, but not on the ends; they are of a good spring temper, and we must keep our gravers *sharp*. After filling in between the leaves with the cement, as recommended—and right here is a point worth mentioning; we should not hold the pinion directly in the flame to heat it, but should fasten it in a light pin vise, and hold this in the flame, allowing the pinion to be heated by conduction—the cement does not fill in up to the full circle of the pinion, but will be festooned, as we might say, from leaf to leaf, as shown in Fig. 2; should any chance to rise above the leaf to interfere with the contact of the leaf with the chuck, it should be removed with a brass scraper.

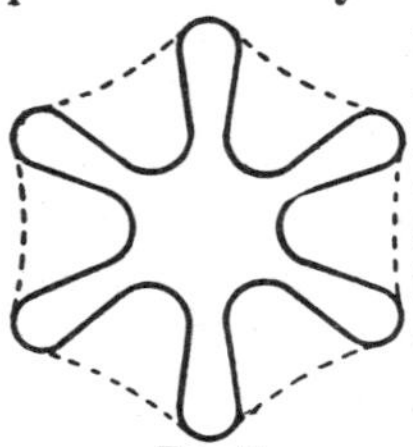

Fig. 2

The pinion should be held in a chuck fitting it perfectly, and running true. The first operation is to turn the seat for the wheel. And before putting the pinion in the chuck, we should observe *about* where this should come; being sure we do not remove so much of the leaves in fitting our wheel that there will be insufficient length of leaf to finish our pinion.

Having settled this point, with a *sharp* lozenge graver, we cut a notch in the pinion, as shown in Fig 3, where *a* represents the face of the chuck; this notch is the beginning of the shoulder on which the wheel is to rest. We then turn the leaves away from toward *b*, slightly taper as shown in dotted lines, until the wheel can be pressed on up to

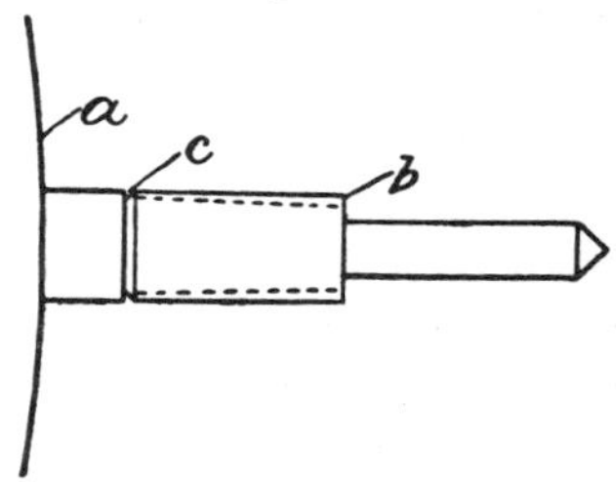

Fig. 3

within its thickness of the shoulder. This method makes it extremely easy to get a perfect fit for the wheel, if we use any care at all.

Of course it is understood that the wheel is first trued out in the hole; this is done before chucking the pinion. We must take care to see how much the hole will bear enlarging, and still afford a sufficient shoulder on the pinion.

We then measure the thickness of the wheel, and measure back from the shoulder, making sufficient allowance for riveting. This will vary with different cases, but generally .3 mm. will be sufficient. After this is determined, our method is to measure up from the wheel seat to locate the shoulder of the pivot, turn and finish the pivot, and reduce the staff to approximately the required size, down to the end, *b*, of the leaves. Our reason for doing this is because the pinion is more rigid, and more easily withstands the stress of turning *before* the leaves are turned off down to the riveting point, than afterward.

We now turn away the leaves down to the riveting point, and with a very sharp graver, reduce the staff to the size required, plus sufficient to grind and polish. We will now form the under cut—this is a trick that frequently gives the beginner trouble; we'll proceed with care—for this we need a very sharp lozenge graver; and it is of advantage to bevel the edge up as shown in Fig. 4, this greatly strengthens the point, besides making the graver more likely to do smooth work, in this particular spot, for reasons which we will explain later. We must not try to jab the graver right in, making it cut on the point and both sides all at once; such procedure brings a great strain on the graver, and is pretty certain to break it. We will, as far as possible, cut one side of the undercut at a time, taking off a light chip first on one side then on the other, gradually deepening it, until it is to the required depth and shape, when a final light shaving from each side, with a very sharp graver, the two cuts meeting at the bottom in

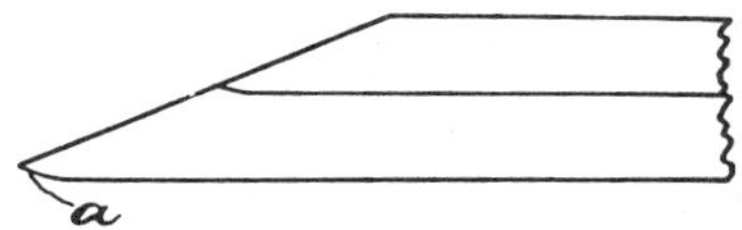

Fig. 4

an angle so perfect and sharp that it can hardly be detected, completes it.

The philosophy of rounding the bottom of the graver for this work, is, besides strengthening the point and edge, it gives it a plane-like quality, as we might call it. Fig. 5 will illustrate this. It will be seen that the rounding, *a*, of the edge may allow the graver to slide along the circle, *b*, without cutting at all, or by tipping the graver slightly, or raising it above the center, it may take a very thin chip, with no possibility of digging in. The rounding must be only slight, and is exaggerated in Fig. 5, for the sake of clearness. After we have the undercut to our satisfaction, we finish the staff, and remove the pinion from the chuck.

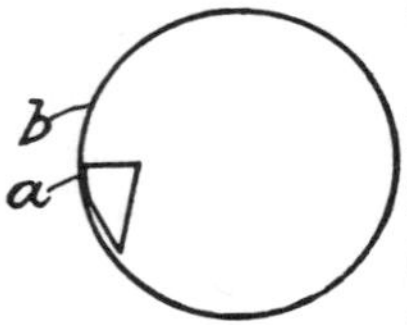

Fig. 5

From this point to the finished pinion, the procedure would vary with different cases; but ordinarily we would reverse the pinion in the chuck, turn the pivot and staff, square off the leaves, turn the undercut, remove the pinion from the chuck, rivet the wheel in place, then return the pinion to the lathe—first removing the cement from between the leaves—holding it by the staff part, and finish the lower pivot, staff, and face the leaves.

It will be seen that in these operations which we perform after riveting the wheel, the pressure is slight, and the pinion may be safely held by the staff part.

Facing pinions with the pivot polisher is very easy if we go at it right. We first grind the face with a cast iron, or soft steel lap—we prefer cast iron—using very fine oilstone powder, then thoroughly clean the job with benzine and watchmakers' putty, (see article on Polishing). We then take a tin lap freshly filed with a fine flat file, and applying a very thin coating to the face of the lap with our steel spatula, we bring the lap *gently* against the end of the leaves, running the lathe at moderate speed—don't overdo it; it is amazing how quickly a beautiful flat black polish is produced, after we have acquired the "knack". *Perfect* facing can be done in this way, so don't give up with anything short of that. The same precautions in the matter of adjusting the lap, belts, etc., must be observed, as in the case of polishing an arbor bearing or pivot, described in the article Bushing.

Our method of measuring, which we devised many years ago, and have published several times, is not generally known and practiced even now. The ordinary method would be to measure from the shoulder of the first completed pivot, up to the point where the other pivot shoulder, and mark or scratch this location on the pinion.

Our method is: suppose the length of the finished pinion between pivot shoulders is 8.3 mm., suppose the new pinion measures from the shoulder of the first pivot made up to the unfinished end, 10 mm.; we then subtract the finished length, from the unfinished and obtain 1.7 mm.; now when the job is grasped in the chuck, we measure down from the unfinished end to locate the shoulder of our pivot. This method is much more certain and accurate, than trying to work to a scratch or other mark. Of course we must not shorten the unfinished end of the pinion any until the shoulder of the pivot is accurately located and the notch cut, then the end of the pinion may be cut off leaving sufficient stock to form the pivot. Study the pinion work in such watches as Patek Philippe, or high grade English watches, and try to do as well, or better.

Staking Safety Pinion Staffs.

Sometimes these staffs slip slightly in the center wheel, causing apparently erratic running of the watch. The pressure which they have to sustain is considerable under ordinary circumstances; and when a mainspring breaks, and the safety pinion sticks—as it frequently does—the staff is very likely to turn in the wheel.

An effective means of making them secure, is to indent the riveting with the special punches, now furnished with nearly all staking tools, for this purpose. The staff should first be properly riveted in the center wheel, by means of smooth punches, just as we would a balance staff.

Then with the indenting punch of proper size, we make four indentations by one suitable blow, and the wheel is secure.

Generally it is better to make four indentions than to try to make eight, for in many cases you will not get the second four exactly between the first four, or exactly the same depth, and the result will not be pleasing; besides, four, when properly made, gives perfect security.

Speaking of safety pinions sticking; many workmen, in cleaning watches, do not remove the pinion head; the result is, in most cases, it rusts fast to the staff. Safety pinions should not only be *removed* when cleaned, but the staff should be *oiled* when the pinion head is replaced.

Using the Micrometer Stop

THE Micrometer Stop is unquestionably the most important and useful adjunct to the staking tool yet devised. It raises the staking tool from the "I guess that's about right" class, to a *precision instrument.* And *more* than this; for it increases the range of usefulness of the staking tool, by making it capable of doing such work as end shaking recessing, etc., with great exactness.

With the Micrometer Stop we may drive a wheel or pinion to pre-determined position *precisely.* We may even measure the thickness of objects, with pefrect accuracy.

Following are instructions for performing some of the operations; the skilful and resourceful workman will readily see other applications.

To drive a new wheel to the exact place occupied by the old one, all that is necessary is to set the pinion in position on the die, and bring down a suitable flat-faced hollow punch in contact with the old wheel. Now slip the stop, Fig. 1, over the top of the punch, bringing the part, *B*, closely in contact with the top of the staking tool; bind the stop to the punch by means of the screw, *E*, operated by handle, *C*, which may be turned in any convenient position by loosening the screw, *D*. Now drive the pinion out of the old wheel (using a different punch) and apply the new wheel; drive the punch until the Stop comes in contact with top of staking tool; the new wheel will then be in the exact place occupied by the old one. This applies to wheels fitting on the smooth arbors of their respective pinions.

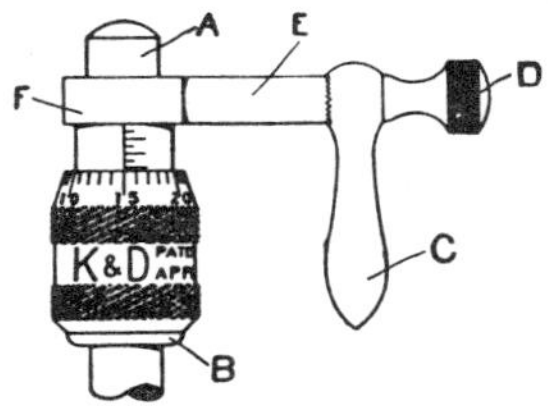

Fig. 1

Now suppose we wish to drive the new wheel to a position farther up or down the pinion than that occupied by the old wheel—and such cases are of frequent occurrence—let us suppose that we have estimated by inspecting the wheel and its action in the watch, that we wish to drive it further down, by an amount equal to one third the wheel's thickness. We measure the wheel; it is, say, .60 mm. thick, therefore we will drive the wheel .20 mm. further down.

We adjust the wheel and pinion on the die—or on a suitable stump—bring down the punch, and adjust the Stop as we did in the first instance— taking care in this case, to see that 0 on the micrometer nut corresponds with the graduation line on the part, *F*,—all being correctly adjusted, we turn the micrometer nut down, through 10 graduations; then drive carefully, until the part, *F*, is in contact with the micrometer nut, and we have driven the wheel exactly .20 mm.

It must be borne in mind that the graduations on the micrometer nut indicate 1–50 mm., therefore, to get the value of the reading in 1–100 mm., we multiply by 2. Having determined a distance of drive in 1–100 mm., we *divide* by 2, and set the Stop accordingly. The Stops are made reading 1–50 mm., to avoid too fine threads, or too fine graduations. Once the workman is familiar with it, he will have no trouble from this source.

The Stop clamps the punch amply tight to be safe against shifting, under any reasonable blow; the operator will, of course exercise discretion, and strike the punch no harder than is necessary to perform the work in hand.

It makes no difference whether the wheel is driven on the pinion, or the wheel rests on the die, or a suitable stump, and the pinion is driven in the wheel. The same principles apply in driving pallet staffs, etc., so they need not be treated in detail.

To get the full limit of drive, the micrometer nut should be run up until its lower edge is about $\frac{1}{8}$ inch above the lower surface of the part, *B*.

To measure the thickness of an object: Suppose we wish to measure the thickness of an escape wheel. We place a small flatfaced stump in the die, and bring down a flat-faced punch upon it, pressing firmly. Now adjust the Stop on top of the punch, taking care that

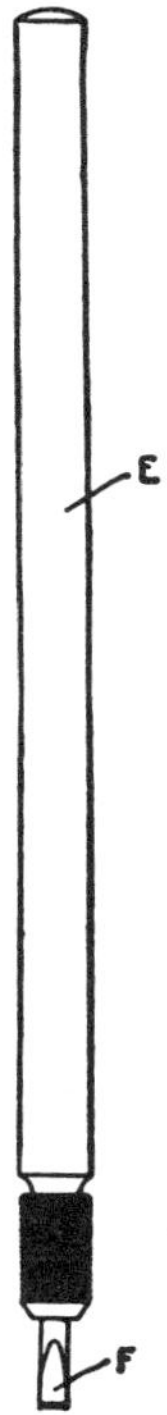

Fig. 2.

0 on the micrometer nut corresponds with the graduation line on the part, *F;* bind the Stop firmly to the punch. Insert the part to be measured between the punch and stump; this raises the part, *F*, above the micrometer nut, a distance equal to the thickness of the object. We now turn the micrometer nut up until it comes in contact with *F*—holding down firmly the while upon the punch—the number of graduations through which we turn the nut, multiplied by 2, gives the thickness of the object in 1–100 mm.

The size of stump employed will be governed to some extent by the character of the object to be measured; some objects may be laid directly on the die.

Any burr that may be around the letters or figures stamped on the upper end of punches, will need to be g ound off before the Micrometer Stop is applied.

For end shaking by milling or other recessing work, we use the cutter spindle shown in Fig. 2, in place of a regular punch. For ordinary end shaking jobs a plain cutter as shown at *F*, Fig. 2, answers admirably, and is very easily made. It is ground flat like a screw driver, then ground to a bevel from both sides like a cold chisel, producing a straight, keen edge.

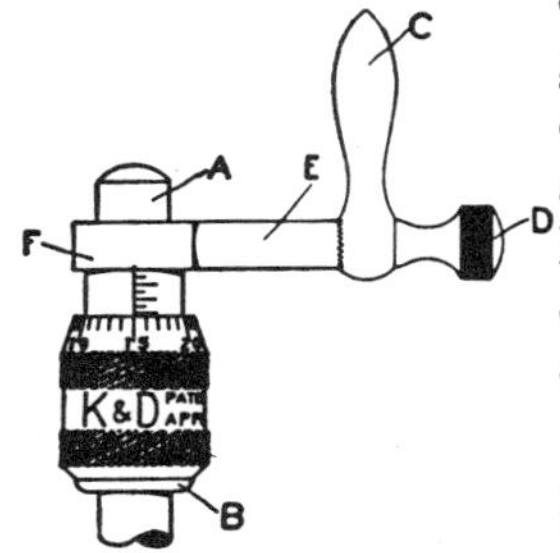

Fig. 3

Adjust the handle to point upward as in Fig. 3. The watch part may be held on a special clamp plate, Fig. 4, which shows its plan and elevation, or it may be held with the fingers directly upon the die of the staking tool; of course the clamp is preferable. Short brass rings of various sizes are provided to lay the watch part upon, when there are projections upon its surface to prevent laying it directly upon the plate. After the work is centered and clamped, we do the milling by turning the cutter spindle by the handle, *C*, at the same time exerting downward pressure.

To set the Stop to allow milling away a given amount, is the same operation substantially as we have described for driving to definite position.

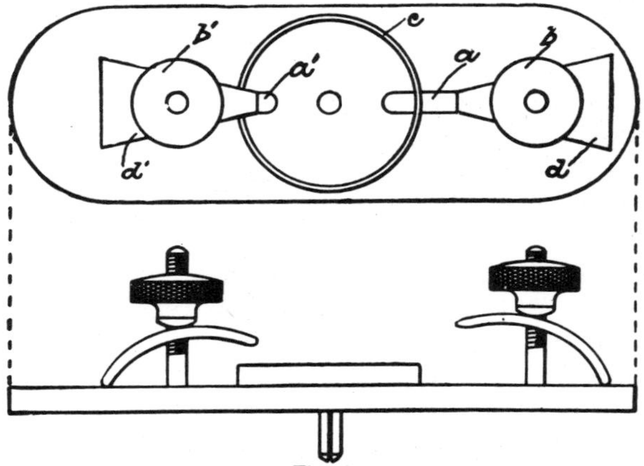

Fig. 4

Regarding Cylinders

CYLINDER pivots are less often broken than the pivots of detached balances. The balances are comparatively light, and the almost constant pressure of an escape wheel tooth against the cylinder, holds the pivots close against the sides of the holes, thus making the effect of a shock—such as dropping the watch—much less disastrous. This is the whole philosophy of placing a doubled piece of paper, or thin sheet metal, under the rim of a balance, when a watch is to be shipped—it holds the pivots close against one side of the holes. To illustrate: when a cowboy tries to ride a bucking broncho, he strives with all his might to stay close to the said broncho; if he succeeds in this, all is well; but if he is thrown up in the air, and in coming down, meets the aforesaid broncho coming up, he is soon demoralized.

Regarding cylinders; the pivots do get broken sometimes, or damaged so badly they have to be replaced. Many workmen pivot cylinders; this may do well enough for the top, but for the bottom we think it much better to make a new plug and pivot entire. This is not at all difficult to the man who can keep his gravers sharp, and turn smoothly. The plug should be turned very slightly taper, squared off, the end finished, and the pivot roughed out, before it is cut off from the wire. If we have a pivot polisher the finishing of the end perfectly square, flat, and highly polished will be easy.

Before going further with making a new plug. it may be well to consider removing the old one. With every staking tool outfit making any pretensions to completeness, are cylinder punches; the small projection on the end of these punches is approximately long enough to force the lower plug out of a cylinder.

Also in every staking tool die of the better class, are two or three chamfered holes, designed for starting cylinder plugs. The proceeding is thus: the lower shell of the cylinder is placed in one of the chamfered holes; with a cylinder punch of suitable size resting on the plug, a light tap on the punch starts it downward as shown in Fig. 1. We now select and center a hole in the die which will just allow the plug to enter, the end of the shell resting upon the flat surface of the die; the plug may now be completely expelled by means of the cylinder punch.

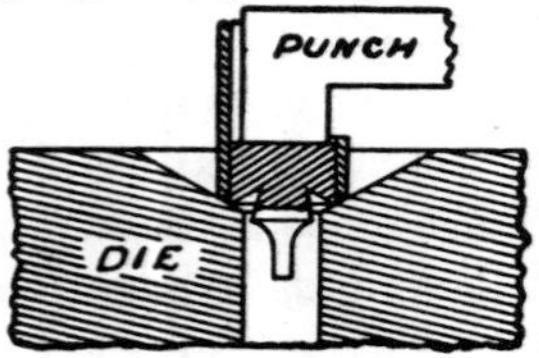

Fig. 1

In inserting the new plug, a cylinder stump is used; these are shaped very much like the punches, but without the short pivot on the end. The cylinder is placed bottom up upon the stump, the point of the stump entering the cylinder and supporting the edge of the lower shell. Now with a suitable flat-faced hollow punch, the plug may be pressed home; the flat top of the stump stopping the plug when precisely even with the top of the lower shell. In our use of the term "lower shell", we refer to the part which encircles the plug.

Now as to finishing the pivot; in these days of true wire chucks and plenty of them, it is usual to push the cylinder out of the brass collet or hub and do the work in a wire chuck. Ordinary staking tool punches are not well adapted to the work of driving the cylinder out, and a special punch must be made. If the workman has the Sub Punch Holder, No. 316 (shown on page 67, catalog section), this is a very easy matter. The punch should be made hollow, with hole large enough to go over the top pivot and rest upon the cylinder. The outer diameter should be slightly less than the cylinder, and the corner slightly rounded, so that in following down through the hub, it will not scrape a chip from one side of the hole.

Many methods have been proposed to get the correct length of a lower cylinder plug from the upper surface to the end of the pivot. A very accurate way is by means of the Micrometer Stop. We would need a sub-punch fitting No. 316, with a long slim pivot turned upon it. Then with the plate held in the special clamp (shown in the article on the Micrometer Stop) with the escape wheel in position, we let the punch down until the long pivot rests upon the lower end stone; we then raise it until its lower end occupies the position with reference to the escape wheel, that the top of the plug should occupy, and read off the distance.

The unfinished pivot may then be stoned off to correct length, or we may measure as recommended in the article Staking Train Pinions.

One of the greatest aids to good work in finishing pivots, we have ever known, is a little sapphire strip, fitted to a suitable holder, as shown in Fig. 2, in which *a*, represents a piece of steel wire about 2 mm. in diameter fitted to a light handle, and having a flat bottomed notch cut for the reception of the sapphire, *b*, which is fastened in position with black lathe wax. The matter of using black wax is important; it imparts a dark look to the sapphire, and enables us to much more readily see where it is cutting when in use. It should be used slightly oiled, and the particles of steel removed will turn the surface grayish; we should keep the sapphire well cleaned off, so we may know exactly where it is cutting. With this little tool it is easily possible to finish pivots quickly to perfect form; it leaves so fine a surface that a light application of an oiled steel burnisher flashes up a mirror-like polish almost instantly.

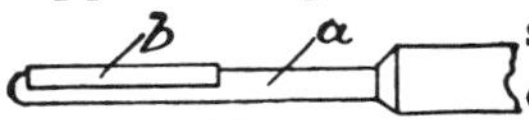

Fig. 2

The strip of sapphire should be about 1.5 cm. long, 2 mm. wide, and 1 mm. thick. It may be obtained of almost any lapidist, and the watchmaker can easily grind it into shape himself with a fine carborundum hone moistened with oil or water. The sapphire should be securely cemented to the edge of a brass block before grinding. This enables one to manage it much better, and to get the curve much truer. The curve should be made of smaller radius than the average cone of balance pivots, then by applying it slightly at an angle, instead of directly crosswise the pivot, the curve may be made to conform to almost any cone. Besides, when applied in this way it produces smoother work, by crossing the grinding lines.

To finish the end of pivots, nothing is so

good and convenient as a small jasper stone, finished flat, like those used to polish jewel settings.

To return a cylinder to the brass collet, the cylinder stump is used to support the upper shell, and a hollow flatfaced punch used to press the collet down. A punch shaped like the cylinder stump is more convenient, in which case the collet is rested on the die, or on a suitable stump. In this case we drive the cylinder down into the collet, instead of driving the collet down onto the cylinder; the work is in plainer view, and better controlled. Such a punch is readily made of a regular cylinder punch by grinding off the pivot.

If we use a cement chuck to hold the cylinders we should clean the cylinder thoroughly with benzine and finally with alcohol, to insure firm adherence of the cement. We should also be sure to make the cement soft enough so it will enter and fill the cylinder; we think the safest way is to fill the cylinder before setting it up in the chuck, using a small strip of copper to apply the cement. The copper may be held in a pin vise, which may be heated instead of the copper. The copper will conduct enough heat to keep the cement fluid, without danger of burning.

The cylinder should be trued by the outside of the shell, when setting it up in the cement.

In returning the cylinder to the brass collet, care must be taken to put it in proper position relative to the banking pin on the rim of the balance.

Our reason for not favoring the plan of pivoting the lower end of a cylinder is: the hole for the plug is nearly always drilled entirely through, throwing up a burr on the inner end of the plug; in addition to this, the pivot plug frequently does not come quite flush with the inner end of the plug; this forms a sort of reservoir which draws the oil away from the surfaces where it is needed; besides this, the job does not look as well.

Enlarging Wheels

ENLARGING train or other wheels, is a job that occasionally comes to the watchmaker; not all the watches we are called upon to repair were right in the first place. The fault of a wheel too small, is most frequently found in Swiss watches of the cheaper class.

Stretching or enlarging a wheel is an operation which we never feel quite satisfied with; it would undoubtedly be much better to fit a new wheel of the proper size, but there's plenty of cases in the experience of every watch repairer when he must resort to such make-shift methods to save his reputation.

He has a cheap watch to repair; it has a faulty depth, perhaps more than one; the owner would be unwilling to pay for new wheels—as a matter of fact the watch is not worth it—if by stretching and rounding up the wheels, in such a case, he makes the watch run satisfactorily, we think he is justified. The wheels were imperfect as they were; and while they are still imperfect, he has made them serve their purpose acceptably; therefore, he has done well.

There can be no question; it's the man who makes a watch run and keep time, who wins and keeps his customers; the quality of work he executes, as regards finish, must be left to his individual judgment, and his conscience. For our own part we believe in making our repairs—as regards finish—fit the job we are working on; in other words we try to preserve the harmonious character of the machine. This seems to us to be the natural and sensible rule; for if we do our work better than the quality of the job calls for, we have created discord that immediately attracts the artistic eye, and the impression is not favorable; the quality of our work may do us credit, but it doesn't fit the job—don't wear your plug hat with overalls.

Having a wheel to stretch, there are several ways of doing it, even with the staking tool, but we shall consider only the methods we follow in our own practice. It is generally necessary to operate on one side of the wheel only, and we, of course, choose the bottom, so any possible marking will not be visible when the watch is set up.

We first select a hole in the die to loosely fit the staff of the pinion, or the entire pinion, as the case may require, and center the part of the rim, upon which we intend to operate, under the punch. It frequently happens that the part of the rim we wish to operate upon comes over, or partly over another hole in the die; when this occurs, we should select a larger hole in the die, and partly plug it with pith, or soft wood, to hold the pinion approximately to the position required; the rim of the wheel should come over a smooth and solid part of the die in every case.

We are now ready to stretch; we may use a flat-faced punch, as large as the width of the rim including the teeth, or we may use a round-faced punch, operating on the middle of the rim only. Either, if carefully used, gives good results, but the flat-faced punch leaves a less conspicuous track.

Having every thing properly adjusted, we tap the punch gently with the hammer, a succession of blows, slowly turning the wheel with a finger, at the same time. If we use the round-faced punch, the visible effect should be a smooth rounded groove along the middle of the rim, instead of a row of dents. If we use the flat-faced punch, the effect is scarcely visible, except where the arms are crossed.

If a wheel has to be stretched very much, we may find it necessary to stretch the arms a little, for the effect of our work on the rim is to make the wheel more or less polygonal. We must guard against this as much as possible; in other words, we must stretch the wheel as round as possible, and not rely upon the rounding up to bring it *round*, but only to the correct size. The nearer we comply with this condition, the more perfect will the finished job be.

Let us examine the reason of this a little. If we take two wheels having the same number of teeth, but one being larger than the other, we know that the distance between the teeth of the larger wheel is greater than it is on the smaller one. Now suppose we take the smaller wheel and stretch the rim only, it will produce high parts between the arms; in other words make the wheel more or less polygonal. At the arms, the only enlargement we get is the spread of the metal in a radial direction; but the spread of the metal under the action of the punch is in the direction of the wheel's circumference as well. Were it not for the arms of the wheel the rim would expand uniformly and remain practically circular; as it is the arms hold the parts of the rim to which they join from spreading, and the rim humps up between the arms—the expanded metal has to go somewhere—the remedy is to stretch the arms separately, enough to bring the wheel circular, and in any case, the rim must not be distorted a great deal, or we will find it difficult to restore the wheel to circular form.

Now suppose we assumed—as many workmen do—that the humping of the rim between the arms was of no consequence, because the rounding-up cutter would restore the wheel to circular form, and so it would, but the distance between the teeth—its circular pitch—would vary, for rounding-up cutters do not correct inequalities of circular pitch, and as we have seen by comparison of the two wheels of different sizes, the teeth of a polygonal wheel—when so made by stretching—are farther apart at the high places than at the end of the arms. And so let us take care to stretch a wheel as near round as possible, and not rely on the cutter to bring it circular.

When we come to using the rounding-up cutter, we should observe several things that are often overlooked. Having selected a cutter of the proper thickness, we adjust the cutter guide snail to deliver the tooth, or rather the space, in such a position that the point of cutter goes exactly in the center of the space. It is not sufficient to simply see that the cutter does not catch; it may not catch, and still come in contact with but one tooth, thus unduly thinning them.

Another point to be looked after carefully, is to mount the wheel in the rounding-up tool so it will turn with perfect freedom, for if it should turn with any considerable degree of stiffness and the cutter should be slightly untrue in the flat, as they often are, the spaces would be cut wider than the thickness of the cutter.

The slight burr produced by the rounding-up cutter may be removed with a fine scratch brush; they may be stoned off by means of a flattened piece of slate pencil, used wet, or the teeth may be smoothed out with charcoal; it is generally preferable to stone off the burr, and smooth off the teeth with charcoal too.

It is hardly feasible to enlarge main-spring barrels, although it is sometimes attempted. A peg may be put through the arbor hole, fitting snugly in one of the holes of the staking tool die; this serves as a pivot or bearing for the barrel to turn on; we then use a flat-faced punch on top of the teeth, or the punch made for stretching balance arms may be used, operating as near the barrel as possible. This process lengthens only one edge of the teeth, and we do not recommend it.

Stretching wheels reminds us that we have several times stretched the balances of cylinder watches to bring them to time. We think it preferable to applying solder to the rim, besides, in many instances there is but scant room for any solder. A very slight enlargement by stretching makes a very considerable difference in the rate of the watch; so if you try it, be careful or you will over-do it. In stretching such a balance, we have always used a flat-faced punch on the bottom of the rim. Of course, the balance should be poised afterward.

Sometimes a cylinder escape wheel needs truing, and the arms are frequently broken in attempts to true them as an ordinary wheel. The arms are easily and safely bent by means of the peening punch. We place the wheel on a brass stump, or a piece of sheet brass on top of a flat faced steel stump, and with the peen punch arranged crosswise the arm we wish to bend, a slight tap on the punch forces the arm down into the brass slightly, and at the same time bends it. Proceed carefully.

TECHNICAL AND REPAIRING - WITH 2002 PRICES

PRICES SUBJECT TO CHANGE WITHOUT NOTICE

A Manual of Veneering - Villard, P. - 174 pp - 5 1/2 x 8 1/2 - 7.95
American Engineer's & Surveyor's Instrument - Gurley Manual 1874 - Garcelon, D. C. - 256 pp - 5 1/2 x 8 1/2 - 19.95
American Watchmaker & Jeweler - Abbott, H. G. - 378 pp - 5 1/2 x 8 1/2 - 19.95
Bench Practices for Watch and ClockmakersSecond Edition-Fried, H. B. - 280 pp - 6 x 9 - PB: 20.00 - HB: 27.50
Berger Hand-book & Illustrated Catalog of the Engineers' & Surveyors' Instruments of Precision, C. L. Berger & Sons - Garcelon, C. 224 pp - 7 x 10 - 22.95
Britten's Watch & Clockmakers Handbook, Dictionary & Guide - 11th Edition - 499 pp - 5 x 9 - HB - 29.50
Building an American Clock Movement - Conover, S. - 84 pp - 8 1/2 x 11 - 21.95
Casting Aluminum - Ammen, C. - 242 pp - 5 1/2 x 8 1/2 - 14.95
Chime Clock Repair - Conover, S. G. - 210 pp - 5 3/4 x 8 3/4 - 28.50
Chip Carving Patterns (Give Patterns for drawer fronts, chair backs, clock faces; many designs & combinations) Barton, W. 128 pp - 7 1/2 x 9 1/2 - 12.95
Classic Clocks for Woodworkers - Complete Patterns for 21 Clocks - Haigh, R. - 160 pp - 8 1/2 x 10 1/4 - PB: 16.95 - HB: 35.00
Clock & Watch Escapements - Gazeley, W .J. - 294 pp - 5 1/2 x 8 3/4 - HB - 29.50
Clock & Watch Escapement - Original Compiled in 1904 by The Keystone (Reprint) - 179 pp - 5 1/2 x 8 1/2 - 12.95
Clock & Watch Pronunciary, The - Gerschler, M. - 254 pp - 5 1/2 x 8 1/2 - 7.95
Clock & Watch Repairing - DeCarle, D. - 309 pp - 5 1/2 x 8 1/2 - HB - 29.50
Clock Repair Basic - Conover, S. G. - 92 pp - 8 1/2 x 11 - 22.95
Clock Repair First Reader, The - Balcomb, P. E. - 180 pp - 8 1/2 x 11 - 16.95
Clock Repair - Part -Time Hours - Full-Time Pay - Pierson, J. R. - 198 pp - 5 1/2 x 8 1/2 - 19.95
Clock Repair Primer - Balcomb, P.E. - 151 pp - 8 1/2 x 11 - 15.95
Clock Repairer's Handbook - Penman, L. - 176 pp - 7 x 10 - PB - 19.95
Clock Repairing & Making - Garrard, F. J. - 168 pp - 5 1/2 x 7 - 22.95

Clockmaking - 18 Antique Designs for the Woodworker - Nelson, J. - 222 pp - 8 1/2 x 10 - HB - 26.95
Clocks & Watches - An Historical Bibliography - Volume 1 - Baillie, G. H. - 414 pp - 6 x 10 - HB - 29.95
Collecting & Repairing Watches - Cutmore, M. - 191 pp - 6 x 9 - HB - 24.95
Compensating Pendulums - Finn, J. L. - 48 pp - 5 1/2 x 8 1/2 - 5.95
Complete Guide to Woodcarving - Hundreds of projects and design. More than 500 drawings and photographs, 683 b/w illustrations - Tangerman, E. J. - 352 pp - 7 x 10 - 17.95
Complete Handbook of Sand Casting - Ammen, C. - 238 pp - 5 x 8 - 13.95
Complete Metalsmith, The - An Illustrated Handbook - McCreight, T. - 150 pp - 6 x 9 - 14.95
Complete Woodfinishing - Hosker, J. - 168 pp - 8 x 10 3/4 - 15.95
Complicated Watches & Their Repair - DeCarle, Donald - 174 pp - 6 1/4 x 9 1/2 - HB - 34.50
Conservation of Clocks and Watches - Wills, P. - 84 pp - 6 x 8 1/2 - HB - 29.95
Engraving on Precious Metals - Brittain & Morton - 228 pp - 5 x 8 - HB - 31.50
Escapements, The - Britton, F.J. - 66 pp - 5 x 8 - 6.95
General Bibliography of Time Measurement - Tardy, L. - 423 pp - 6 x 9 - 89.50
Gold Leaf Techniques - LeBlanc - 3rd Edition - 165 pp - 7 x 10 - HB - 32.95
Glory of Gold - The - A Contemporary Approach to Gilding - Trytell, S. - 80 pp - 8 1/2 x 11 1/4 - HB - 24.95
Gluing & Clamping - A Woodworker's Handbook - Spielman, P. - 256 pp - 8 x 10 - 15.95
Hairsprings - Manipulating of Old & New Springs - 45 pp - 5 x 8 - 5.95
Helpful Information for Watchmakers - 11th Edition - Ohlson, O. - 104 pp - 5 x 8 - 11.95
Hobson's Choice - English Bracket Clock Repairing - Hobson, C. -101 pp - 8 x 12 - 26.95
Horological Hints & Helps - Britten, F.W. - 375 pp - 5 x 7 - HB - 29.50
Horolovar 400 Day Repair Guide - Terwilliger, C. - 10 th Edition - 227 pp - 9 x 10 - HB - 24.95
How to Build a Regulator clock - Huckabee, J. - 80 pp - 8 x 11 - 18.00
How to Cast Small Metal & Rubber Parts - 2nd Edition - Cannon, W. -168 pp - 5 x 8 - 14.95
How to Make an Electric Clock - Way, R. B. - 53 pp - 5 x 7 - 8.00
How to Make Tiny Drills quickly, easily, & accurately - Porter, R. - 16 pp - 5 x 8 - 3.95
How to Manage and Regulate Clocks and Watches - Berthoud, M. F. - 81 pp - 5 1/2 x 8 1/4 - 18.00

How to Repair Herschede Tubular Bell Clocks - Conover, S. G. - 116 pp - 8 x 11 - 28.50
How to Repair Musical Boxes - Jacot & Son - 32 pp - 5 x 8 - 4.95
Instructions for Installation and Maintenance of Self-Winding Synchronized Clocks see Tran Duy Ly's "American Clocks" Volume 2.
Jewelry Engraver's Manual, The - Revised Edition - Hardy, R. - 143 pp - 6 x 9 - Original: 9.95
Jewelry Repair Manual - Hardy, R.A. - 2nd Edition - 253 pp - 6 x 9 - 12.95
Juanelo Turriano - Charles V's Clockmaker Man & His Legend, The - Garcia - Diego, J.A. 165 pp - 7 x 9 - HB - 35.00
Les Brocot-une dynastie d'horlogers - Chavigny, R. - 224 pp - 6 x 9 - HB - 69.95
Living On Air - History of The Atmos Clock - Lebet, J. - 88 pp - 6 1/2 x 9 1/4 - 20.00
Making an Eight Day Longcase Clock - Timmins, A. - 104 pp - 8 x 11 - HB - 39.95
Making Clocks - Penman, L. - 117 pages - 7 x 10 1/4 - 35.00
Making Modern Jewellery - Bagley, P. - 128 pp - 8 1/2 x 10 1/4 - HB - 24.95
Making & Repairing Wooden Clock Cases - Taylor, V. & Babb, H. - 191 pp - 8 x 11 - HB - 29.95
Making Wooden Clock Cases - Designs, Plans & Instructions for 20 Clocks - Tim & Ashby, P. - 222 pp - 9 x 11 3/4 - 24.95
Manual for Overhaul, Repair & Handling of Hamilton Model 21 Ship Chronometer with Parts Catalog with Addition of May 1st 1946 Hamilton Marine Chronometer Repair Parts List - US Navy Department - 104 pp - 8 x 11 - 17.95
Manual for Overhaul, Repair & Handling of Hamilton 35-Size Chronometer Watch with Parts Catalog-US Navy Department 66 pp - 8 1/2 x 11 - 9.95
Manual for Overhaul, Repair, & Handling of U. S. Navy Mechanical, Boat & Deck Clocks Chelsea Type with Parts Catalog US Navy Department - 108 pp - 8 1/2 x 11 - 17.95
Mechanical Singing-bird Tabatieres - Mayson, G. - This book is a complete record of the development, construction and restoration of these miniature mechanical marvels and is beautifully with over 200 color illustrations. 256 pp - 7/34 x 10 - HB - 195.00
Metalwork and Enamelling - 5th Revised Edition - Maryon, H. - 335 pp - 5 x 9 - 9.95
Metalwork for Craftsmen - Kronquist, E. - 202 pp - 5 x 8 - 8.95
Modern Clock - Construction Regulating & Repair - Goodrich, W. (This is one of the best books on the market for repairing American and European clocks; including Electric, 400-Day and Cuckoo clocks.) 1997 Edition - 502 pp - 6 x 9 - HB - 27.50
My Own Right Time - An Exploration of Clockwork Design - Woodward, P. - 166 pp - 8 x 10 - HB - 55.00
New & Tradition Styles of Chip Carving - From Classic to Positive Imaging - Barton, W. - 128 pp - 7 1/2 X 9 1/2 - 12.95

New Haven Clocks & Watches - With a Special Section on New Haven Movements Ly, T. D. - Released in 1997, this book does justice to the vast line of clocks and watches produced by one of America's most prolific makers. The 50 pages of movements and parts showing extremely clear and detailed illustrations of 61 New Haven movements are outstanding. With 2000 Price Update - 520 pp - 8 1/2 x 11 - Over 2,200 quality illustrations and photographs. - Smyth sewn hardbound on white gloss enamel text. 55.00 - See Book Review

Novelty Clocks for Woodworkers - Haigh, R. - 128 pp - 8 1/2 x 10 1/2 - PB: 16.95 - HB: 24.95

Pendulum Clock or Geometrical Demonstration Concerning the Motion of Pendula as Applied to Clocks - Huygens, C. - 182 pp - 6 x 9 - HB - 25.00

Pin-Pallet Escapement - Adjustment & Repair, The - Porter, R. - 43 pp - 8 x 11 - 14.95

Practical Balance & Hairspring Work - Kleinlein, W. J. - 115 pp - 5 x 8 - 11.95

Practical Benchwork for Horologist - Levin, L. & S. - Revised - Eighth Edition - 381 pp - 6 x 9 - 19.95

Practical Casting - A Studio Reference - McCreight, T. - Revised Edtion - 153 pp - 6 x 9 - 19.95

Practical Clock Repairing - DeCarle, D. - 244 pp - 5 1/2 x 8 3/4 - HB - 35.00

Practical Jewelry Repair - Hickling, J. - 198 pp - 5 1/2 x 8 3/4 - HB - 29.50

Practical Watch Repairing - DeCarle, D. - 319 pp - 5 1/2 x 8 3/4 - HB - 35.00

Precision Time Measures: Their Construction & Repair - Higginbotham, C. T. - 343 pp - 5 x 8 - HB - 27.50

Proper Use of Watchmaker's Graver - Barkus, H. - 37 pp - 5 x 8 - 5.00

Quartz Movement Reference Manual - 92 pp - 5 x 8 - 18.95

Questions & Answers in Quartz Watch Repairing - Carpenter, A. - 106 pp - 8 1/2 x 11 - 12.95

Questions and Answers on Watch and Clock Making - Elgin Watchmakers College - 40 pp - 8 1/2 x 11 - 7.95

Repairing Antique Clocks - Smith, E. - 232 pp - 5 3/4 x 8 3/4 - HB - 24.95

Repairing Old Clocks and Watches - Whiten, A. - 280 pp - 5 1/2 x 8 3/4 - HB - 34.50

Repairing Watch Cases - Schwanatus and Fenimore - 34 pp - 5 x 8 - 5.95

Restoring Grandfather Clocks - Smith, B. & E. - 235 pp - 6 1/2 x 9 1/4 - HB - 40.00

Revere Clocks Service Manual - Welch, T. - 24 pp - 5 x 8 - 6.00

Scroll Saw Fretwork Patterns - Spielman & Reidle - Over 200 full-sized patterns - 256 pp - 8 x 10 - 14.95

Seth Thomas Clocks and Movements 2nd Edition is now out of print. A revised and enlarged 3rd Edition on prepairing.

Spiral Hairspring, with a new evaluation of its elastic properties - Woodward, P. - 56 pp - 5 1/2 x 8 - 10.95

Square Recoil Escapement - Aydlett, G. - 35 pp - 6 x 9 - 5.50

Staff Making & Pivoting - Hall, E. - 46 pp - 5 x 7 - 5.95
Staking Tools & How to Use Them - K & D Manufacturing Corp. - 108 pp - 5 x 8 - 12.95
Striking Clocks - Baier, J.G. - 51 pp - 5 x 8 - 7.95
Striking Clock Repair Guide - Conover, S. G. - 90 pp - 8 1/2 x 11 - 22.95
Techniques in Reverse Glass Painting - Emery, M. - 129 pp - 7 x 8 - 39.95
Timetrax - Clock Train Tables - 1996 Edition - 26 pp - 5 1/2 x 8 1/2 - 6.00
Timing Manipulations - A.W.I. - 63 pp - 6 x 9 - 4.50
Top 300 Trade Secret of a Master Clockmaker, The - Huckabee, J. M. - 111 pp - 8 1/2 x 11 - 26.95
Tower Clock & How to Make It - Ferson, E. - 57 pp - 5 x 7 - 6.00
Turning Lathes : A Guide to Turning, Screw Cutting, Metal Spinning and Ornamental Turning - Lukin, J. - 432 pp - 6 x 9 1/4 - 24.95
Twenty-First Century Watchmaking - Book 1 - Smith, W. - 212 pp - 7 x 10 - HB - 45.00
Victorian Scroll Saw Patterns - Complete instruction with close-up drawings and photos. - Spielman, P. - 192 pp - 8 x 10 - 14.95
Watch & Clock Encyclopedia - DeCarle, D. - 326 pp - 6 x 9 - 3rd Edition - HB - 37.50
Watch & Clock Information, Please - Samelius & Hagens - 408 pp - 6 x 9 - HB - 30.00
Watch & Clock Making & Repairing - Gazeley, W. J. - 452 pp - 5 1/2 x 8 1/2 - HB - 37.50
Watch Escapement - Fried, H. B. - 191 pp - 5 x 8 - 17.95
Watch Repairer's Manual - Fried, H. B. - 4th Edition - 453 pp - 6 x 9 - HB - 35.00
Watch Repairing As a Hobby - Fletcher, D. W. - 64 pp - 5 x 7 - 8.95
Watch Repairing, Cleaning & Adjusting - Garrard, F. J. - 222 pp - 5 x 8 - 16.95
Watchmaker's Lathe - Goodrich, W. - 272 pp - 6 x 9 - HB - Very good - 21.95
Watchmaker's Models Engineer's Lathe - A User's Manual - De Carle, D. - 4th Edi.1985 - Revised by Restabb, E. - 195 pp - 6 x 8 - HB - 29.95
Wheel & Pinion Cutting in Horology-Wild, M. - 253 pp - 8 x 10 - HB - 55.00
Working in Precious Metals - Smith, E. - 400 pp - 5 x 8 - HB - 31.95

TOOLS - WITH 2002 PRICES

PRICES SUBJECT TO CHANGE WITHOUT NOTICE

A Price Guide To Antique Tools - Kean, H. P. & Pollak, E. S. - 184 pp - 6 x 9 - 14.95
American Machinist's Tools - An Illustrated Directory of Patents - Over 1000 Diagrams - Cope, K. L. - 420 pp - 7 x 10 - 29.95
Antique Tools, Our American Heritage with Price Guide - McNerney, K. - 156 pp - 5 x 8 - 9.95
Art of Welding, The - Vause, W. A. - 95 pp - 6 x 8 - 14.95
Care & Operation of a Lathe, The - Sheldon Machine Co. - 108 pp - 5 x 8 - 11.95
Clock Wheel & Pinion Cutting-Wild, J. M. - 48 pp - 6 x 8 - 12.95
Collecting Antique Tools - Kean, H. P. & Pollak, E. S. - 203 pp - 8 1/2 x 11 - 29.95
Dies - Their Construction and Use - Woodworth, J. V. - 400 pp - 5 1/2 x 8 1/2 - (Reprint) 15.95
Drilling and Thread Cutting Handbook No. 1 - 1925 - Popular Mechanics - 94 pp - 6 x 9 - (Reprint) 7.95
Electroplating - Reetz, H. C. - (Reprint) - 99 pp - 4 1/2 x 8 - 7.95
507 Mechanical Movements - Brown, H. T. - 128 pp - 6 x 7 - (Reprint) 7.95
Foundrywork for The Amateur - Aspin, B. - 94 pp - 6 x 8 - 15.95
Gear Cutting Practice - Colvin & Stanley - 344 pp - 5 1/2 x 8 1/2 - (Reprint) 14.95
Getting The Most Out of Your Lathe - Published by Delta - 48 pp - 6 x 9 - 5.95
Green-Sand Casting - 174 pp - 5 1/2 x 8 1/2 - (Reprint) - 9.95
Handbook of Mechanical Design - Nordenholt, Kerr & Sasso - 277 pp - 81/2 x 11 - (Reprint) 19.95
Hardening, Tempering, Annealing and Forging of Steel - Woodworth, J. V. - 288 pp - 5 1/2 x 8 1/2 - (Reprint) 9.95
Home Machinist's Handbook, The - A Complete do-it-yourselfer's Guide to Machine work... with projects - Briney, D. - 275 pp - 7 x 9 - 19.95
How to Run a Lathe - 1942 - South Bend Lathe Works - 128 pp - 5 1/2 x 8 1/2 - (Reprint) 7.95
Keep Your Lathe in Trim - South Bend Lathe Works - 28 pp - 5 1/2 x 8 1/2 - (Reprint) - 3.95
Lathe and Shapening Machine Tools - Duplex - 4 3/4 x 7 - 11.95
Lathe Design - Construction and Operation - Perrigo, O. - 469 pp - 5 1/2 x 8 1/2 - (Reprint) 24.95
Lathe Handbook No. 1 - 1925 - Popular Mechanics - 87 pp - 6 x 9 - (Reprint) 7.95

Lathe operations - Barritt, J. W. - 176 pp - 8 1/2 x 11 - 13.95
Lapp & Flershem-1887-1888 Catalog & Price List of Tools, Materials & Findings for Watchmakers & Jewlers - 280 pp - 7 x 10 - 17.95
Making Small Workshop Tools - Bray, S. - 94 pp - 6 x 8 - 15.95
Measuring and Marking Metals - Law, I. - 110 pp - 6 x 8 - 14.95
Metal Turner's Handybook - Hasluck, P. - 144 pp - 4 x 7 - 22.95
Metal Working Tools - Bradley, I. - 105 pp - 5 x 7 - 9.95
Metalcaster's Bible - Ammen, C. - 434 pp - 5 x 8 - 21.95
Metalworking - Tools, Material, and Processes - Hasluck, P. N. - 760 pp - 6 x 9 - HB - (Reprint) - 29.95
Milling in Small Lathe - Marshall, P. - 64 pp - 4 3/4 x 7 - 10.95
Milling Machine Operations - Barritt, J. W. - 110 pp - 8 1/2 x 11 - (Reprint) 9.95
Milling Machine Practices - The Cincinnati Milling Machine Co. - 28 pp - 6 x 9 - 3.95
Milling Operations in the Lathe - Cain, T. - 125 pp - 6 x 8 - 15.95
Power Tools: Care & Use - Veasey, W. - 64 pp - 6 x 9 - 7.95
Practical Lessons in Metal Turning & Screwcutting - Marshall, P. - 178 pp - 4 x 7 - 22.95
Practical Metal Plate Work - Hasluck, P. N. - 160 pp - 5 1/2 x 8 1/2 - (Reprint) - 9.95
Proper Use of the Watchmaker's Graver - Barkus, H.A. - 37 pp - 5 x 8 - 5.00
Restoring, Turning & Using Classic Woodworking Tools - Duncan, M. - 256 pp - 7 3/4 x 10 - 14.95
Running a Milling Machine - Colvin, F. H. - 157 pp - 5 1/2 x 8 1/2 - (Reprint) - 7.50
Running an Engine Lathe - Colvin, F. H. - 117 pp - 5 1/2 x 8 1/2 - 6.95
Saw & Sawings - Bradley, I. - 95 pp - 6 x 8 - 14.95
Shaper Operations - Barritt, J. W. - 55 pp - 8 1/2 x 11 - (Reprint) 7.95
Sheet Metal Work - Wakeford, R. - 150 pp - 6 x 8 - 15.95
Soldering and Brazing - Yates, R. F. - 90 pp - 5 1/2 x 8 1/2 - (Reprint) 6.95
South Bend Lathe Booklets - South Bend Lathe Works - 96 pp - 6 x 9 - (Reprint) 7.95
Staking Tools & How to Use Them - K & D Manufacturing Corporation - 108 pp - 5 1/2 x 8 1/2 - 12.95
Tools and Their Uses - U.S. Navy - 179 pp - 6 x 9 - 5.95
Treatise on Milling and Milling Machines - Cincinnati Milling Machine Co. - 409 pp - 5 1/2 x 8 1/2 - (Reprint) - 13.95

Turning And Mechanical Manipulation - Holtzapffel's - Vol. I: Preparation & Various Modes of Working Them - 462 pp - 6 x 9 1/4 - 24.95
Turning And Mechanical Manipulation - Holtzapffel's - Vol. II : Construction, Action and Application of Cutting Tools - 1025 pp - 6 x 9 1/4 - 22.95
Turning And Mechanical Manipulation - Holtzapffel's - Vol. III: Abrasive, Process Not Accomplished With Cutting Tools - 796 pp - 6 x 9 1/4 - 32.50
Turning Lathes: A Guide to Turning, Screw Cutting, Metal Spinning and Ornamental Turning - Lukin, J. - 432 pp - 6 x 9 1/4 - 24.95
Turning & Milling in Horology - Nicolet, J. - 56 pp - 8 x 10 - HB - 44.95
Watchmaker's Lathe - Goodrich, W. - 272 pp - 6 x 9 - HB - Very good - 21.95
Watchmaker's Model Engineer's Lathe - A User's Manual - De Carle, D. - 4th Edition 1985, Revised by Restabb, E. - 195 pp - 6 x 8 - HB - 29.95
Watchmakers' Staking Tool, The - Lucchina, G. & Perkins, A. - 118 pp - 6 x 9 - HB - 24.95
Watch-makers' Tools, Material and Supplies - Hoffman Supply Co.- 162 pp - 6 x 9 (Reprint) - 9.95
Welding & Cutting - Oats, J. - 92 pp - 5 x 7 - 9.95
Wheels & Pinions & How to Determine Their Exact Size - Shouffelberger, H. & Gribi, T. 20 pp - 8 x 11 - 6.00
Workholding in the Lathe - Cain, T. - 112 pp - 5 x 8 - 15.95
Workshop Drawing - Cain, T. - 110 pp - 6 x 8 - 14.95